HOW TO
FLY A HORSE

HOW TO
FLY A HORSE

THE SECRET HISTORY OF CREATION,

INVENTION AND DISCOVERY

KEVIN ASHTON

125 YEARS

WILLIAM HEINEMANN: LONDON

1 3 5 7 9 10 8 6 4 2

William Heinemann
20 Vauxhall Bridge Road
London SW1V 2SA

William Heinemann is part of the Penguin Random House group of
companies whose addresses can be found at global.penguinrandomhouse.com.

Penguin
Random House
UK

First published by William Heinemann in 2015
(First published in the United States by Doubleday,
a division of Random House LLC, New York in 2015)

www.randomhouse.co.uk

A CIP catalogue record for this book is available from the British Library.

ISBN 9780434022908 (Hardback)
ISBN 9780434022915 (Trade paperback)

Book design by Pei Loi Koay
Jacket illustration by Christoph Niemann

Printed and bound in Great Britain by Clays Ltd, St Ives plc

Penguin Random House is committed to a sustainable future for our business,
our readers and our planet. This book is made from Forest Stewardship
Council® certified paper.

A genius is the one most like himself.

— THELONIOUS MONK

Work your best at being you. That's where home is.

—BILL MURRAY

CONTENTS

PREFACE: THE MYTH

In 1815, Germany's *General Music Journal* published a letter in which Mozart described his creative process:

> When I am, as it were, completely myself, entirely alone,
> and of good cheer; say traveling in a carriage, or walking after
> a good meal, or during the night when I cannot sleep; it is on
> such occasions that my ideas flow best and most abundantly. All
> this fires my soul, and provided I am not disturbed, my subject
> enlarges itself, becomes methodized and defined, and the whole,
> though it be long, stands almost finished and complete in my
> mind, so that I can survey it, like a fine picture or a beautiful
> statue, at a glance. Nor do I hear in my imagination the parts
> successively, but I hear them, as it were, all at once. When I
> proceed to write down my ideas the committing to paper is
> done quickly enough, for everything is, as I said before, already
> finished; and it rarely differs on paper from what it was in my
> imagination.

In other words, Mozart's greatest symphonies, concertos, and operas came to him complete when he was alone and in a good mood. He needed no tools to compose them. Once he had finished imagining his masterpieces, all he had to do was write them down.

This letter has been used to explain creation many times. Parts of it appear in *The Mathematician's Mind,* written by Jacques Hadamard in 1945, in *Creativity: Selected Readings,* edited by Philip Vernon in 1976, in Roger Penrose's award-winning 1989 book, *The Emperor's New Mind,* and it is alluded to in Jonah Lehrer's 2012 bestseller *Imagine.* It influenced the poets Pushkin and Goethe and the playwright Peter Shaffer. Directly and indirectly, it helped shape common beliefs about creating.

But there is a problem. Mozart did not write this letter. It is a forgery. This was first shown in 1856 by Mozart's biographer Otto Jahn and has been confirmed by other scholars since.

Mozart's real letters—to his father, to his sister, and to others—reveal his true creative process. He was exceptionally talented, but he did not write by magic. He sketched his compositions, revised them, and sometimes got stuck. He could not work without a piano or harpsichord. He would set work aside and return to it later. He considered theory and craft while writing, and he thought a lot about rhythm, melody, and harmony. Even though his talent and a lifetime of practice made him fast and fluent, his work was exactly that: work. Masterpieces did not come to him complete in uninterrupted streams of imagination, nor without an instrument, nor did he write them down whole and unchanged. The letter is not only forged, it is false.

It lives on because it appeals to romantic prejudices about invention. There is a myth about how something new comes to be. Geniuses have dramatic moments of insight where great things and thoughts are born whole. Poems are written in dreams. Symphonies are composed complete. Science is accomplished with eureka shrieks. Businesses are built by magic touch. Something is not, then is. We do not see the road from nothing to new, and maybe we do not want to. Artistry must be misty magic, not sweat and grind. It dulls the luster to think that every elegant equation, beautiful painting, and brilliant machine is born of effort and error, the progeny of false starts and failures, and that each maker is as flawed, small, and mortal as the rest of us. It is seductive to conclude that great innovation is delivered to us by miracle via genius. And so the myth.

The myth has shaped how we think about creating for as long as creating has been thought about. In ancient civilizations, people believed that things could be discovered but not created. For them, everything had *already* been created; they shared the perspective of Carl Sagan's joke on this topic: "If you want to make an apple pie from scratch, you must first invent the universe." In the Middle Ages, creation was possible but was reserved for divinity and those with divine inspiration. In the Renaissance, humans were finally thought capable of creation, but they had to be *great men*—Leonardo, Michelangelo, Botticelli, and the like. As the nineteenth century turned into the twentieth, creating became a subject for philosophical, then psychological investigation. The question being investigated was "How do the great men do it?" and the answer had the residue of medieval divine intervention. A lot of the meat of the myth was added at this time, with the same few anecdotes about epiphanies and genius—including hoaxes like Mozart's letter—being circulated and recirculated. In 1926, Alfred North Whitehead made a noun from a verb and gave the myth its name: *creativity*.

The creativity myth implies that few people can be creative, that any successful creator will experience dramatic flashes of insight, and that creating is more like magic than work. A rare few have what it takes, and for them it comes easy. Anybody else's creative efforts are doomed.

How to Fly a Horse is about why the myth is wrong.

I believed the myth until 1999. My early career—at London University's student newspaper, at a Bloomsbury noodle start-up called Wagamama, and at a soap and paper company called Procter &

Gamble—suggested that I was not good at creating. I struggled to execute my ideas. When I tried, people got angry. When I succeeded, they forgot that the idea was mine. I read every book I could find about creation, and each one said the same thing: ideas come magically, people greet them warmly, and creators are winners. My ideas came gradually, people greeted them with heat instead of warmth, and I felt like a loser. My performance reviews were bad. I was always in danger of being fired. I could not understand why my creative experiences were not like the ones in the books.

It first occurred to me that the books might be wrong in 1997, when I was trying to solve an apparently boring problem that turned out to be interesting. I could not keep a popular shade of Procter & Gamble lipstick on store shelves. Half of all stores were out of stock at any given time. After much research, I discovered that the cause of the problem was insufficient information. The only way to see what was on a shelf at any moment was to go look. This was a fundamental limit of twentieth-century information technology. Almost all the data entered into computers in the 1900s came from people typing on keyboards or, sometimes, scanning bar codes. Store workers did not have time to stare at shelves all day, then enter data about what they saw, so every store's computer system was blind. Shopkeepers did not discover that my lipstick was out of stock; shoppers did. The shoppers shrugged and picked a different one, in which case I probably lost the sale, or they did not buy lipstick at all, in which case the store lost the sale, too. The missing lipstick was one of the world's smallest problems, but it was a symptom of one of the world's *biggest* problems: computers were brains without senses.

This was so obvious that few people noticed it. Computers were fifty years old in 1997. Most people had grown up with them and had grown used to how they worked. Computers processed data that people entered. As their name confirmed, computers were regarded as thinking machines, not sensing machines.

But this is not how intelligent machines were originally conceived. In 1950, Alan Turing, computing's inventor, wrote, "Machines will eventually compete with men in all purely intellectual fields. But which are the best ones to start with? Many people think that a very abstract activity, like the playing of chess, would be best. It can also be maintained that it is best to provide the machine with the best sense organs that money can buy. Both approaches should be tried."

Yet few people tried that second approach. In the twentieth century, computers got faster and smaller and were connected together, but they did not get "the best sense organs that money can buy." They did not get any "sense organs" at all. And so in May 1997, a computer called Deep Blue could beat the reigning human chess world champion, Garry Kasparov, for the first time ever, but there was no way a computer could see if a lipstick was on a shelf. This was the problem I wanted to solve.

I put a tiny radio microchip into a lipstick and an antenna into a shelf; this, under the catchall name "Storage System," became my first patented invention. The microchip saved money and memory by connecting to the Internet, newly public in the 1990s, and saving its data there. To help Procter & Gamble executives understand this system for connecting things like lipstick—and diapers, laundry detergent, potato chips, or any other object—to the Internet, I gave it a short and ungrammatical name: "the Internet of Things." To help make it real, I started working with Sanjay Sarma, David Brock, and Sunny Siu at the Massachusetts Institute of Technology. In 1999, we cofounded a research center, and I emigrated from England to the United States to become its executive director.

In 2003, our research had 103 corporate sponsors, plus additional labs in universities in Australia, China, England, Japan, and Switzerland, and the Massachusetts Institute of Technology signed a lucrative license deal to make our technology commercially available.

In 2013, my phrase "Internet of Things" was added to the Oxford

Dictionaries, which defined it as "a proposed development of the Internet in which everyday objects have network connectivity, allowing them to send and receive data."

Nothing about this experience resembled the stories in the "creativity" books I had read. There was no magic, and there had been few flashes of inspiration—just tens of thousands of hours of work. Building the Internet of Things was slow and hard, fraught with politics, infested with mistakes, unconnected to grand plans or strategies. I learned to succeed by learning to fail. I learned to expect conflict. I learned not to be surprised by adversity but to prepare for it.

I used what I discovered to help build technology businesses. One was named one of the ten "Most Innovative Companies in the Internet of Things" in 2014, and two were sold to bigger companies—one less than a year after I started it.

I also gave talks about my experiences of creating. My most popular talk attracted so many people with so many questions that, each time I gave it, I had to plan to stay for at least an hour afterward to answer questions from audience members. That talk is the foundation of this book. Each chapter tells the true story of a creative person; each story comes from a different place, time, and creative field and highlights an important insight about creating. There are tales within the tales, and departures into science, history, and philosophy.

Taken together, the stories reveal a pattern for how humans make new things, one that is both encouraging and challenging. The encouraging part is that everyone can create, and we can show that fairly conclusively. The challenging part is that there is no magic moment of creation. Creators spend almost all their time creating, persevering despite doubt, failure, ridicule, and rejection until they succeed in making something new and useful. There are no tricks, shortcuts, or get-creative-quick schemes. The process is ordinary, even if the outcome is not.

Creating is not magic but work.

HOW TO
FLY A HORSE

CHAPTER I

CREATING IS ORDINARY

1 | EDMOND

In the Indian Ocean, fifteen hundred miles east of Africa and four
thousand miles west of Australia, lies an island that the Portuguese
knew as Santa Apolónia, the British as Bourbon, and the French, for
a time, as Île Bonaparte. Today it is called Réunion. A bronze statue
stands in Sainte-Suzanne, one of Réunion's oldest towns. It shows an
African boy in 1841, dressed as if for church, in a single-breasted jacket,
bow tie, and flat-front pants that gather on the ground. He wears no
shoes. He holds out his right hand, not in greeting but with his thumb
and fingers coiled against his palm, perhaps about to flip a coin. He is
twelve years old, an orphan and a slave, and his name is Edmond.

The world has few statues of Africa's enslaved children. To understand why Edmond stands here, on this lonely ocean speck, his hand
held just so, we must travel west and back, thousands of miles and hundreds of years.

On Mexico's Gulf Coast, the people of Papantla have dried the fruit of a vinelike orchid and used it as a spice for more millennia than they remember. In 1400, the Aztecs took it as tax and called it "black flower." In 1519, the Spanish introduced it to Europe and called it "little pod," or *vainilla*. In 1703, French botanist Charles Plumier renamed it "vanilla."

Vanilla is hard to farm. Vanilla orchids are great creeping plants, not at all like the *Phalaenopsis* flowers we put in our homes. They can live for centuries and grow large, sometimes covering thousands of square feet or climbing five stories high. It has been said that lady's slippers are the tallest orchids and tigers the most massive, but vanilla dwarfs them both. For thousands of years, its flower was a secret known only to the people who grew it. It is not black, as the Aztecs were led to believe, but a pale tube that blooms once a year and dies in a morning. If a flower is pollinated, it produces a long, green, beanlike capsule that takes nine months to ripen. It must be picked at precisely the right time. Too soon and it will be too small; too late and it will split and spoil. Picked beans are left in the sun for days, until they stop ripening. They do not smell of vanilla yet. That aroma develops during curing: two weeks on wool blankets outdoors each day before being wrapped to sweat each night. Then the beans are dried for four months and finished by hand with straightening and massage. The result is oily black lashes worth their weight in silver or gold.

Vanilla captivated the Europeans. Anne of Austria, daughter of Spain's King Philip III, drank it in hot chocolate. Queen Elizabeth I of England ate it in puddings. King Henry IV of France made adulterating it a criminal offense punishable by a beating. Thomas Jefferson discovered it in Paris and wrote America's first recipe for vanilla ice cream.

But no one outside Mexico could make it grow. For three hundred years, vines transported to Europe would not flower. It was only in 1806 that vanilla first bloomed in a London greenhouse and three more decades before a plant in Belgium bore Europe's first fruit.

The missing ingredient was whatever pollinated the orchid in the wild. The flower in London was a chance occurrence. The fruit in Belgium came from complicated artificial pollination. It was not until late in the nineteenth century that Charles Darwin inferred that a Mexican insect must be vanilla's pollinator, and not until late in the twentieth century that the insect was identified as a glossy green bee called *Euglossa viridissima*. Without the pollinator, Europe had a problem. Demand for vanilla was increasing, but Mexico was producing only one or two tons a year. The Europeans needed another source of supply. The Spanish hoped vanilla would thrive in the Philippines. The Dutch planted it in Java. The British sent it to India. All attempts failed.

This is where Edmond enters. He was born in Sainte-Suzanne in 1829. At that time Réunion was called Bourbon. His mother, Mélise, died in childbirth. He did not know his father. Slaves did not have last names—he was simply "Edmond." When Edmond was a few years old, his owner, Elvire Bellier-Beaumont, gave him to her brother Ferréol in nearby Belle-Vue. Ferréol owned a plantation. Edmond grew up following Ferréol Bellier-Beaumont around the estate, learning about its fruits, vegetables, and flowers, including one of its oddities—a vanilla vine Ferréol had kept alive since 1822.

Like all the vanilla on Réunion, Ferréol's vine was sterile. French colonists had been trying to grow the plant on the island since 1819. After a few false starts—some orchids were the wrong species, some soon died—they eventually had a hundred live vines. But Réunion saw no more success with vanilla than Europe's other colonies had. The orchids seldom flowered and never bore fruit.

Then, one morning late in 1841, as the spring of the Southern Hemisphere came to the island, Ferréol took his customary walk with Edmond and was surprised to find two green capsules hanging from the vine. His orchid, barren for twenty years, had fruit. What came next surprised him even more. Twelve-year-old Edmond said he had pollinated the plant himself.

To this day there are people in Réunion who do not believe it. It seems impossible to them that a child, a slave, and, above all, an *African,* could have solved the problem that beat Europe for hundreds of years. They say it was an accident—that he was trying to damage the flowers after an argument with Ferréol or he was busy seducing a girl in the gardens when it happened.

Ferréol did not believe the boy at first. But when more fruit appeared, days later, he asked for a demonstration. Edmond pulled back the lip of a vanilla flower and, using a toothpick-sized piece of bamboo to lift the part that prevents self-fertilization, he gently pinched its pollen-bearing anther and pollen-receiving stigma together. Today the French call this *le geste d'Edmond*—Edmond's gesture. Ferréol called the other plantation owners together, and soon Edmond was traveling the island teaching other slaves how to pollinate vanilla orchids. After seven years, Réunion's annual production was a hundred pounds of dried vanilla pods. After ten years, it was two tons. By the end of the century, it was two *hundred* tons and had surpassed the output of Mexico.

Ferréol freed Edmond in June 1848, six months before most of Réunion's other slaves. Edmond was given the last name Albius, the Latin word for "whiter." Some suspect this was a compliment in racially charged Réunion. Others think it was an insult from the naming registry. Whatever the intention, things went badly. Edmond left the plantation for the city and was imprisoned for theft. Ferréol was unable to prevent the incarceration but succeeded in getting Edmond released after three years instead of five. Edmond died in 1880, at the age of fifty-one. A small story in a Réunion newspaper, *Le Moniteur,* described it as a "destitute and miserable end."

Edmond's innovation spread to Mauritius, the Seychelles, and the huge island to Réunion's west, Madagascar. Madagascar has a perfect environment for vanilla. By the twentieth century, it was producing most of the world's vanilla, with a crop that in some years was worth more than $100 million.

The demand for vanilla increased with the supply. Today it is the world's most popular spice and, after saffron, the second most expensive. It has become an ingredient in thousands of things, some obvious, some not. Over a third of the world's ice cream is Jefferson's original flavor, vanilla. Vanilla is the principal flavoring in Coke, and the Coca-Cola Company is said to be the world's largest vanilla buyer. The fine fragrances Chanel No. 5, Opium, and Angel use the world's most expensive vanilla, worth $10,000 a pound. Most chocolate contains vanilla. So do many cleaning products, beauty products, and candles. In 1841, on the day of Edmond's demonstration to Ferréol, the world produced fewer than two thousand vanilla beans, all in Mexico, all the result of pollination by bees. On the same day in 2010, the world produced more than five million vanilla beans, in countries including Indonesia, China, and Kenya, almost all of them—including the ones grown in Mexico—the result of *le geste d'Edmond*.

2 | COUNTING CREATORS

What is unusual about Edmond's story is not that a young slave created something important but that he got the credit for it. Ferréol worked hard to ensure that Edmond was remembered. He told Réunion's plantation owners that it was Edmond who first pollinated vanilla. He lobbied on Edmond's behalf, saying, "This young negro deserves recognition from this country. It owes him a debt, for starting up a new industry with a fabulous product." When Jean Michel Claude Richard, director of Réunion's botanical gardens, said he had developed the technique and shown it to Edmond, Ferréol intervened. "Through old age, faulty memory or some other cause," he wrote, "Mr. Richard now imagines that he himself discovered the secret of how to pollinate vanilla, and imagines that he taught the technique to the person who discovered it! Let us leave him to his fantasies." Without Ferréol's great effort, the truth would have been lost.

In most cases, the truth *has* been lost. We do not know, for example, who first realized that the fruit of an orchid could be cured until it tastes good. Vanilla is an innovation inherited from people long forgotten. This is not exceptional; it is normal. Most of our world is made of innovations inherited from people long forgotten—not people who were rare but people who were common.

Before the Renaissance, concepts like authorship, inventorship, or claiming credit barely existed. Until the early fifteenth century, "author" meant "father," from the Latin word for "master," *auctor. Auctor*-ship implied authority, something that, in most of the world, had been the divine right of kings and religious leaders since Gilgamesh ruled Uruk four thousand years earlier. It was not to be shared with mere mortals. An "inventor," from *invenire,* "find," was a discoverer, not a creator, until the 1550s. "Credit," from *credo,* "trust," did not mean "acknowledgment" until the late sixteenth century.

This is one reason we know so little about who made what before the late 1300s. It is not that no records were made—writing has been around for millennia. Nor is it that there was no creation—everything we use today has roots stretching back to the beginning of humanity. The problem is that, until the Renaissance, people who created things didn't matter much. The idea that at least *some* people who create things should be recognized was a big step forward. It is why we know that Johannes Gutenberg invented printing in Germany in 1440 but not who invented windmills in England in 1185, and that Giunta Pisano painted the crucifix in Bologna's Basilica of San Domenico in 1250 but not who made the mosaic of Saint Demetrios in Kiev's Golden-Domed Monastery in 1110.

There are exceptions. We know the names of hundreds of ancient Greek philosophers, from Acrion to Zeno, as well as a few Greek engineers of the same period, such as Eupalinos, Philo, and Ctesibius. We also know of a number of Chinese artists from around 400 C.E. onward, including the calligrapher Wei Shuo and her student Wang Xizhi. But the general principle holds. Broadly speaking, our knowledge of who created what started around the middle of the thirteenth century, increased during the European Renaissance of the fourteenth to seventeenth centuries, and has kept increasing ever since. The reasons for the change are complicated and the subject of debate among historians—they include power struggles within the churches of Europe, the rise of science, and the rediscovery of ancient philosophy—but there is little doubt that most creators started getting credit for their creations only after the year 1200.

One way this happened was through patents, which give credit within rigorous constraints. The first patents were issued in Italy in the fifteenth century, in Britain and the United States in the seventeenth century, and in France in the eighteenth century. The modern U.S. Patent and Trademark Office granted its first patent on July 31, 1790. It granted its *eight millionth* patent on August 16, 2011. The patent office does not keep records of how many different people have been granted patents, but economist Manuel Trajtenberg developed a way of working it out. He analyzed names phonetically and compared matches with zip codes, coinventors, and other information to identify each unique inventor. Trajtenberg's data suggests that more than six million distinct individuals had received U.S. patents by the end of 2011.

The inventors are not distributed evenly across the years. Their numbers are increasing. The first million inventors took 130 years to get their patents, the second million 35 years, the third million 22 years, the fourth million 17 years, the fifth million 10 years, and the sixth million inventors took 8 years. Even with foreign inventors removed and adjustments for population increase, the trend is unmistakable. In 1800, about one in every 175,000 Americans was granted a first patent. In 2000, one in every 4,000 Americans received one.

Not all creations get a patent. Books, songs, plays, movies, and other works of art are protected by copyright instead, which in the United States is managed by the Copyright Office, part of the Library of Congress. Copyrights show the same growth as patents. In 1870, 5,600 works were registered for copyright. In 1886, the number grew to more than 31,000, and Ainsworth Spofford, the librarian of Congress, had to plead for more space. "Again it becomes necessary to refer to the difficulty and embarrassment of prosecuting the annual enumeration of the books and pamphlets recently completed," he wrote in a report to Congress. "Each year and each month adds to the painfully overcrowded condition of the collections, and although many rooms have been filled with the overflow from the main Library, the difficulty of handling so large an accumulation of unshelved books is constantly growing." This became a refrain. In 1946, register of copyrights Sam Bass Warner reported that "the number of registrations of copyright claims rose to 202,144 the greatest number in the history of the Copyright Office, and a number so far beyond the capacities of the existing staff that Congress, responding to the need, generously provided for additional personnel." In 1991, copyright registrations reached a peak of more than 600,000. As with patents, the increase exceeded population growth. In 1870, there was 1 copyright registration for every 7,000 U.S. citizens. In 1991, there was one copyright registration for every 400 U.S. citizens.

More credit is given for creation in science, too. The *Science Citation Index* tracks the world's leading peer-reviewed journals in science and technology. For 1955, the index lists 125,000 new scientific papers—about 1 for every 1,350 U.S. citizens. For 2005, it lists more than 1,250,000 scientific papers—one for every 250 U.S. citizens.

Patents, copyrights, and peer-reviewed papers are imperfect proxies. Their growth is driven by money as well as knowledge. Not all work that gets this recognition is necessarily good. And, as we shall see later, giving credit to individuals is misleading. Creation is a chain reaction: thousands of people contribute, most of them anonymous, all of them creative. But, with numbers so big, and even though we miscount and undercount, the point is hard to miss: over the last few centuries, more people from more fields have been getting more credit as creators.

We have not become more creative. The people of the Renaissance were born into a world enriched by tens of thousands of years of human invention: clothes, cathedrals, mathematics, writing, art, agriculture, ships, roads, pets, houses, bread, and beer, to name a fraction. The second half of the twentieth century and the first decades of the twenty-first century may appear to be a time of unprecedented innovation, but there are other reasons for this, and we will discuss them later. What the numbers show is something else: when we start counting creators, we find that a lot of people create. In 2011, almost as many Americans received their first patent as attended a typical NASCAR race. Creating is not for an elite few. It is not even *close* to being for an elite few.

The question is not whether invention is the sole province of a tiny minority but the opposite: how many of us are creative? The answer, hidden in plain sight, is all of us. Resistance to the possibility that Edmond, a boy with no formal education, could create something important is grounded in the myth that creating is extraordinary. Creating is not extraordinary, even if its results sometimes are. Creation is human. It is all of us. It is everybody.

3 | THE SPECIES OF NEW

Even without numbers, it is easy to see that creation is not the exclusive domain of rare geniuses with occasional inspiration. Creation surrounds us. Everything we see and feel is a result of it or has been touched by it. There is too much creation for creating to be infrequent.

This book is creation. You probably heard about it via creation, or the person who told you about it did. It was written using creation, and creation is one reason you can understand it. You are either lit by creation now or you will be, come sundown. You are heated or cooled or at least insulated by creation—by clothes and walls and windows. The sky above you is softened by fumes and smog in the day and polluted by electric light at night—all results of creation. Watch, and it will be crossed by an airplane or a satellite or the slow dissolve of a vapor trail. Apples, cows, and all other things agricultural, apparently natural, are also creation: the result of tens of thousands of years of innovation in trading, breeding, feeding, farming, and—unless you live on the farm—preservation and transportation.

You are a result of creation. It helped your parents meet. It likely assisted your birth, gestation, and maybe conception. Before you were born, it eradicated diseases and dangers that could have killed you. After, it inoculated and protected you against others. It treated the illnesses you caught. It helps heal your wounds and relieve your pain. It did the same for your parents and their parents. It recently cleaned you, fed you, and quenched your thirst. It is why you are where you are. Cars, shoes, saddles, or ships transported you, your parents, or your grandparents to the place you now call home, which was less habitable before creation—too hot in the summer or too cold in the winter or too wet or too swampy or too far from potable water or freely growing food or prowled by predators or all of the above.

Listen, and you hear creation. It is in the sound of passing sirens; distant music; church bells; cell phones; lawn mowers and snow blowers; basketballs and bicycles; waves on breakers; hammers and saws; the creak and crackle of melting ice cubes; even the bark of a dog, a wolf changed by millennia of selective breeding by humans; or the purr of a cat, the descendant of one of just five African wildcats that humans have been selectively breeding for ten thousand years. Anything that is as it is due to conscious human intervention is invention, creation, new.

Creation is so around and inside us that we cannot look without seeing it or listen without hearing it. As a result, we do not notice it at all. We live in symbiosis with new. It is not something we do; it is something we are. It affects our life expectancy, our height and weight and gait, our way of life, where we live, and the things we think and do. We change our technology, and our technology changes us. This is true for every human being on the planet. It has been true for two thousand generations; ever since the moment our species started thinking about improving its tools.

Anything we create is a tool—a fabrication with purpose. There is nothing special about species with tools. Beavers make dams. Birds build nests. Dolphins use sponges to hunt for fish. Chimpanzees use sticks to dig for roots and stone hammers to open hard-shelled food. Otters use rocks to break open crabs. Elephants repel flies by making branches into switches they wave with their trunks. Clearly our tools are better. The Hoover Dam beats the beaver dam. But why?

Our tools have not been better for long. Six million years ago, evolution forked. One path led to chimpanzees—distant relatives, but the closest living ones we have. The other path led to us. Unknown numbers of human species emerged. There was *Homo habilis, Homo heidelbergensis, Homo ergaster, Homo rudolfensis,* and many others, some whose status is still controversial, some still to be discovered. All human. None us.

Like other species, these humans used tools. The earliest were pointed stones used to cut nuts, fruit, and maybe meat. Later, some human species made two-sided hand axes requiring careful masonry and nearly perfect symmetry. But apart from minor adjustments, human tools were monotonous for a million years, unchanged no matter when or where they were used, passed through twenty-five thousand generations without modification. Despite the mental focus needed to make it, the design of that early human hand ax, like the design of a beaver dam or bird's nest, came from instinct, not thought.

Humans that looked like us first appeared 200,000 years ago. This was the species called *Homo sapiens*. Members of *Homo sapiens* did not act like us in one important way: their tools were simple and did not change. We do not know why. Their brains were the same size as ours. They had our opposable thumbs, our senses, and our strength. Yet for 150,000 years, like the other human species of their time, they made nothing new.

Then, 50,000 years ago, something happened. The crude, barely recognizable stone tools *Homo sapiens* had been using began to change—and change quickly. Until this moment, this species, like all other animals, did not innovate. Their tools were the same as their parents' tools and their grandparents' tools and their great-grandparents' tools. They made them, but they didn't make them better. The tools were inherited, instinctive, and immutable—products of evolution, not conscious creation.

Then came by far the most important moment in human history— the day one member of the species looked at a tool and thought, "I can make this better." The descendants of this individual are called *Homo sapiens sapiens*. They are our ancestors. They are us. What the human race created was creation itself.

The ability to change anything was the change that changed everything. The urge to make better tools gave us a massive advantage over all other species, including rival species of humans. Within a few tens of thousands of years, all other humans were extinct, displaced by an anatomically similar species with only one important difference: ever-improving technology.

What makes our species different and dominant is innovation. What is special about us is not the size of our brains, speech, or the mere fact that we use tools. It is that each of us is in our own way driven to make things better. We occupy the evolutionary niche of new. The niche of new is not the property of a privileged few. It is what makes humans human.

We do not know exactly what evolutionary spark caused the ignition of innovation 50,000 years ago. It left no trace in the fossil record. We do know that our bodies, including our brain size, did not change—our immediate pre-innovation ancestor, *Homo sapiens,* looked exactly like us. That makes the prime suspect our mind: the precise arrangement of, and connections between, our brain cells. Something structural seems to have changed there—perhaps as a result of 150,000 years of fine-tuning. Whatever it was, it had profound implications, and today it lives on in everyone. Behavioral neurologist Richard Caselli says, "Despite great qualitative and quantitative differences between individuals, the neurobiologic principles of creative behavior are the same from the least to the most creative among us." Put simply, we all have creative minds.

This is one reason the creativity myth is so terribly wrong. Creating is not rare. We are all born to do it. If it seems magical, it is because it is innate. If it seems like some of us are better at it than others, that is because it is part of being human, like talking or walking. We are not all equally creative, just as we are not all equally gifted orators or athletes. But we can all create.

The human race's creative power is distributed in all of us, not concentrated in some of us. Our creations are too great and too numerous to come from a few steps by a few people. They must come from many steps by many people. Invention is incremental—a series of slight and constant changes. Some changes open doors to new worlds of opportunity and we call them breakthroughs. Others are marginal. But when we look carefully, we will always find one small change leading to another, sometimes within one mind, often among several, sometimes across continents or between generations, sometimes taking hours or days and occasionally centuries, the baton of innovation passing in an endless relay of renewal. Creating accretes and compounds, and as a consequence, every day, each human life is made possible by the sum of all previous human creations. Every object in our life, however old or new, however apparently humble or simple, holds the stories, thoughts, and courage of thousands of people, some living, most dead—the accumulated new of fifty thousand years. Our tools and art are our humanity, our inheritance, and the everlasting legacy of our ancestors. The things we make are the speech of our species: stories of triumph, courage, and creation, of optimism, adaptation, and hope; tales not of one person here and there but of one people everywhere; written in a common language, not African, American, Asian, or European but human.

There are many beautiful things about creating being human and innate. One is that we all create in more or less the same way. Our individual strengths and tendencies of course cause differences, but they are small and few relative to the similarities, which are great and many. We are more like Leonardo, Mozart, and Einstein than not.

4 | AN END TO GENIUS

The Renaissance belief that creating is reserved for genius survived through the Enlightenment of the seventeenth century, the Romanticism of the eighteenth century, and the Industrial Revolution of the nineteenth century. It was not until the middle of the twentieth century that the alternative position—that everyone is capable of creation—first emerged from early studies of the brain.

In the 1940s, the brain was an enigma. The body's secrets had been revealed by several centuries of medicine, but the brain, producing consciousness without moving parts, remained a puzzle. Here is one reason theories of creation resorted to magic: the brain, throne of creation, was three pounds of gray and impenetrable mystery.

As the West recovered from World War II, new technologies appeared. One was the computer. This mechanical mind made understanding the brain seem possible for the first time. In 1952, Ross Ashby synthesized the excitement in a book called *Design for a Brain*. He summarized the new thinking elegantly:

> The most fundamental facts are that the earth is over
> 2,000,000,000 years old and that natural selection has
> been winnowing the living organisms incessantly. As a result they
> are today highly specialized in the arts of survival, and among
> these arts has been the development of a brain, an organ that has
> been developed in evolution as a specialized means to survival.
> The nervous system, and living matter in general, will be assumed
> to be essentially similar to all other matter. No *deus ex machina* will
> be invoked.

Put simply: brains don't need magic.

A San Franciscan named Allen Newell came of academic age during this period. Drawn by the energy of the era, he abandoned his plan to become a forest ranger (in part because his first job was feeding gangrenous calves' livers to fingerling trout), became a scientist instead, and then, one Friday afternoon in November 1954, experienced what he would later call a "conversion experience" during a seminar on mechanical pattern recognition. He decided to devote his life to a single scientific question: "How can the human mind occur in the physical universe?"

"We now know that the world is governed by physics," he explained, "and we now understand the way biology nestles comfortably within that. The issue is how does the mind do that as well? The answer must have the details. I've got to know how the gears clank, how the pistons go and all of that."

As he embarked on this work, Newell became one of the first people to realize that creating did not require genius. In a 1959 paper called "The Processes of Creative Thinking," he reviewed what little psychological data there was about creative work, then set out his radical idea: "Creative thinking is simply a special kind of problem-solving behavior." He made the point in the understated language academics use when they know they are onto something:

> The data currently available about the processes involved in creative and non-creative thinking show no particular differences between the two. It is impossible to distinguish, by looking at the statistics describing the processes, the highly skilled practitioner from the rank amateur. Creative activity appears simply to be a special class of problem-solving activity characterized by novelty, unconventionality, persistence, and difficulty in problem formulation.

It was the beginning of the end for genius and creation. Making intelligent machines forced new rigor on the study of thought. The capacity to create was starting to look more and more like an innate function of the human brain—possible with standard equipment, no genius necessary.

Newell did not claim that everyone was equally creative. Creating, like any human ability, comes in a spectrum of competence. But everybody can do it. There is no electric fence between those who can create and those who cannot, with genius on one side and the general population on the other.

Newell's work, along with the work of others in the artificial intelligence community, undermined the myth of creativity. As a result, some of the next generation of scientists started to think about creation differently. One of the most important of these was Robert Weisberg, a cognitive psychologist at Philadelphia's Temple University.

Weisberg was an undergraduate during the first years of the artificial intelligence revolution, spending the early 1960s in New York before getting his PhD from Princeton and joining the faculty at Temple in 1967. He spent his career proving that creating is innate, ordinary, and for everybody.

Weisberg's view is simple. He builds on Newell's contention that creative thinking is the same as problem solving, then extends it to say that creative thinking is the same as thinking in general but with a creative result. In Weisberg's words, "when one says of someone that he or she is 'thinking creatively,' one is commenting on the outcome of the process, not on the process itself. Although the impact of creative ideas and products can sometimes be profound, the mechanisms through which an innovation comes about can be very ordinary."

Said another way, normal thinking is rich and complex—so rich and complex that it can sometimes yield extraordinary—or "creative"—results. We do not need other processes. Weisberg shows this in two ways: with carefully designed experiments and detailed case studies of creative acts—from the painting of Picasso's *Guernica* to the discovery of DNA and the music of Billie Holiday. In each example, by using a combination of experiment and history, Weisberg demonstrates how creating can be explained without resorting to genius and great leaps of the imagination.

Weisberg has not written about Edmond, but his theory works for Edmond's story. At first, Edmond's discovery of how to pollinate vanilla came from nowhere and seemed miraculous. But toward the end of his life, Ferréol Bellier-Beaumont revealed how the young slave solved the mystery of the black flower.

Ferréol began his story in 1793, when German naturalist Konrad Sprengel discovered that plants reproduced sexually. Sprengel called it "the secret of nature." The secret was not well received. Sprengel's peers did not want to hear that flowers had a sex life. His findings spread anyway, especially among botanists and farmers who were more interested in growing good plants than in judging floral morality. And so Ferréol knew how to manually fertilize watermelon, by "marrying the male and female parts together." He showed this to Edmond, who, as Ferréol described it, later "realized that the vanilla flower also had male and female elements, and worked out for himself how to join them together." Edmond's discovery, despite its huge economic impact, was an incremental step. It is no less creative as a result. All great discoveries, even ones that look like transforming leaps, are short hops.

Weisberg's work, with subtitles like *Genius and Other Myths* and *Beyond the Myth of Genius,* did not eliminate the magical view of creation nor the idea that people who create are a breed apart. It is easier to sell secrets. Titles available in today's bookstores include *10 Things Nobody Told You About Being Creative, 39 Keys to Creativity, 52 Ways to Get and Keep Your Creativity Flowing, 62 Exercises to Unlock Your Most Creative Ideas, 100 What-Ifs of Creativity,* and *250 Exercises to Wake Up Your Brain.* Weisberg's books are out of print. The myth of creativity does not die easily.

But it is becoming less fashionable, and Weisberg is not the only expert advocating for an epiphany-free, everybody-can theory of creation. Ken Robinson was awarded a knighthood for his work on creation and education and is known for the moving, funny talks he gives at an annual conference in California called TED (for technology, entertainment, and design). One of his themes is how education suppresses creation. He describes "the really extraordinary capacity that children have, their capacity for innovation," and says that "all kids have tremendous talents and we squander them, pretty ruthlessly." Robinson's conclusion is that "creativity now is as important in education as literacy, and we should treat it with the same status." Cartoonist Hugh MacLeod makes the same point more colorfully: "Everyone is born creative; everyone is given a box of crayons in kindergarten. Being suddenly hit years later with the 'creative bug' is just a wee voice telling you, 'I'd like my crayons back, please.' "

If genius is a prerequisite for creating, it should be possible to identify creative ability in advance. The experiment has been tried many times. The best-known version was started in 1921 by Lewis Terman and still continues. Terman, a cognitive psychologist born in the nineteenth century, was a eugenicist who believed the human race could be improved with selective breeding, a classifier of individuals according to their abilities as he perceived them. His most famous classification system was the Stanford-Binet IQ test, which placed children on a scale "ranging from idiocy on the one hand to genius on the other," with classifications in between including "retarded," "feeble-minded," "delinquent," "dull normal," "average," "superior," and "very superior." Terman was so sure of his test's accuracy that he thought its results revealed immutable destiny. He also believed, like all eugenicists, that African Americans, Mexicans, and others were genetically inferior to English-speaking white people. He described them as "the world's hewers of wood and drawers of water" who lacked the ability to be "intelligent voters or capable citizens." The children, he said, "should be segregated in special classes." The adults should "not be allowed to reproduce." *Unlike* almost all eugenicists, Terman set out to prove his prejudices.

His experiment was called Genetic Studies of Genius. It was a longitudinal study—meaning it would follow its subjects for a long period of time. It tracked more than 1,500 children who lived in California, all of whom were identified as "gifted" by Terman's IQ test or some similar scheme. Nearly all the participants were white and from upper- or middle-class families. The majority of them were male. This is unsurprising: of the 168,000 children considered for that pool of 1,500, only one was black, one was Indian, one was Mexican, and four were Japanese. The selectees, who had an average IQ of 151, called themselves "Termites." Data about the progress of their lives were collected every five years. After Terman died, in 1956, others took up his research, aiming to continue the work until the last participant either withdrew or died.

Thirty-five years into the experiment, Terman proudly enumerated the success of "his children":

> Nearly 2,000 scientific and technical papers and articles and
> some 60 books and monographs in the sciences, literature, arts,
> and humanities. Patents granted amount to at least 230. Other
> writings include 33 novels, about 375 short stories, novelettes, and
> plays; 60 or more essays, critiques, and sketches; and 265 miscellaneous articles. Hundreds of publications by journalists that classify as news stories, editorials, or newspaper columns. Hundreds,
> if not thousands, of radio, television, or motion picture scripts.

The identity of most of the Termites is confidential. Around thirty have disclosed their participation. Some were notable creators. Jess Oppenheimer worked in television and was a principal developer of a top-ranked, Emmy Award–winning comedy called *I Love Lucy*. Edward Dmytryk was a film director, making more than fifty Hollywood movies, including *The Caine Mutiny*, which was nominated for several Oscars, starred Humphrey Bogart, and was the second most watched film of 1954.

Other participants fared less well. They found more ordinary work as policemen, technicians, truck drivers, and secretaries. One was a potter who was eventually committed to a mental hospital; another cleaned swimming pools; several collected welfare. By 1947, Terman was forced to conclude, "We have seen that intellect and achievement are far from perfectly correlated." This was despite Terman actively helping his participants by writing letters of recommendation and providing mentorship and references. Movie director Dmytryk benefited from a letter at age fourteen, after he ran away from his violent father. Terman explained to the Los Angeles juvenile authorities that Dmytryk was "gifted" and his case deserved special consideration. He was saved from his abusive childhood and placed into a good foster home. TV producer Oppenheimer was a coat salesman until Terman helped him get into Stanford University. Some Termites landed careers in Terman's field of educational psychology, and many were admitted to Stanford, where he was an eminent professor. One Termite took over the study after Terman died.

The study's flaws and biases are beside the point. What matters is what happened to the children Terman excluded. The genius theory of creating predicts that the only creators among the children will be the ones Terman deemed geniuses. None of those excluded should have done anything creative: after all, they were not geniuses.

This is where Terman's study falls flat. Terman did not create a control group of non-geniuses for comparison. We know a lot about the hundreds of children who were selected and only a little about the tens of thousands who were not. But what we do know is sufficient to undermine the genius theory. One child Terman considered and rejected was a boy named William Shockley. Another was a boy named Luis Alvarez. Both grew up to win Nobel Prizes for physics— Shockley for coinventing the transistor, Alvarez for his work in nuclear magnetic resonance. Shockley started Shockley Semiconductor, one of the first electronics companies in Silicon Valley. Employees of Shockley's went on to found Fairchild Semiconductor, Intel, and Advanced Micro Devices. Working with his son Walter, Alvarez was the first to propose that an asteroid caused the extinction of the dinosaurs—the "Alvarez hypothesis"—which, after decades of controversy, scientists now accept as fact.

Terman's failure to identify these innovators does not close the coffin on the genius hypothesis. Perhaps his definition of genius was insufficient or Shockley and Alvarez's tests were wrongly administered. But the magnitude of their achievements begs us to consider another conclusion: genius does not predict creative ability because it is not a prerequisite.

Subsequent studies tried to correct this by measuring creative ability specifically. Starting in 1958, psychologist Ellis Paul Torrance administered a set of tests later known as the Torrance Tests of Creative Thinking to schoolchildren in Minnesota. Tasks included coming up with unusual ways to use a brick, having ideas for improving a toy, and improvising a drawing based on a given shape, such as a triangle. The researchers assessed the creative ability of each child by looking at how many ideas he or she generated, how different the ideas were from the others, how unusual they were, and how much detail they included. The difference in thinking about thinking that characterized psychology after World War II is evident in Torrance's work. Torrance suspected that creation was "within the reach of everyday people in everyday life" and eventually tried to modify his tests to eliminate racial and socioeconomic bias. Unlike Terman, Torrance did not expect his method to be a reliable predictor of future outcomes. "A high degree of these abilities does not guarantee that the possessor will behave in a highly creative manner," he wrote. "A high level of these abilities, however, increases a person's chances of behaving creatively."

How did these more modest expectations play out for Torrance's Minnesotan children? The first follow-up research came in 1966, using children who were tested in 1959. They were asked to select the three classmates who had the best ideas and then complete a questionnaire about their own creative work. The answers were compared with the data from seven years earlier. The correlation was not bad. It was certainly better than Terman's. The results were much the same after a second follow-up test, in 1971. The Torrance Tests seemed to be a reasonable way to predict creative ability.

The moment of truth came after fifty years, when the participants were ending their careers and had demonstrated whatever creative ability they possessed. The results were simple. Sixty participants responded. None of the high-scoring individuals had created anything that had achieved public recognition. Many had done things Torrance and his followers called "personal achievements" of creation, such as forming an action group, building a house, or pursuing a creative hobby. The Torrance Tests had achieved the modest goal of predicting who might have a somewhat creative life. They had done nothing to foresee who might have a creative career.

Without meaning to, Torrance had done something else. He had reinforced what Terman's results showed but Terman stubbornly ignored: that genius has nothing to do with creative ability, even when creative ability is broadly defined and generously measured. Torrance had recorded the IQ of all his participants. His results showed no connection between creative ability and general intelligence. Whatever Terman was measuring had nothing to do with creating, which is why he missed the Nobel laureates Shockley and Alvarez. We may call them creative geniuses now, but if creative genius is apparent only after creation, it is just another way of saying "creative."

6 | ORDINARY ACTS

The case against genius is clear: too many creators, too many creations, and too little predetermination. So how does creation happen?

The answer lies in the stories of people who have created things. Stories of creation follow a path. Creation is destination, the consequence of acts that appear inconsequential by themselves but that, when accumulated, change the world. Creating is an ordinary act, creation its extraordinary outcome.

Was Edmond's story ordinary or extraordinary? If we could travel back to Ferréol's estate in the Réunion of 1841, we would see ordinary acts: a boy following an old man around a garden, a conversation about watermelons, the boy poking around inside a flower. If we returned in

1899, we would see an extraordinary outcome: the island transformed, the world transforming. Knowing the outcome tempts us to retrofit the acts with extraordinariness—to picture Edmond awake all night wrestling with the problem of pollination, having a moment of epiphany in the moonlight, and an enslaved twelve-year-old orphan revolutionizing Réunion and the world.

But creation comes from *ordinary* acts. Edmond learned about botany through boyish curiosity and daily walks with Ferréol. Ferréol kept up with developments in the science of plants, including the work of Charles Darwin and Konrad Sprengel. Edmond applied this knowledge to vanilla, with the help of a bamboo tool and a child's small fingers. When we look behind creation's curtain, we find people like us doing things we can do.

This does not make creating easy. Magic is instant, genius an accident of birth. Take them away and what is left is work.

Work is the soul of creation. Work is getting up early and going home late, turning down dates and giving up weekends, writing and rewriting, reviewing and revising, rote and routine, staring down the doubt of the blank page, beginning when we do not know where to start, and not stopping when we cannot go on. It is not fun, romantic, or, most of the time, even interesting. If we want to create, we must, in the words of Paul Gallico, open our veins and bleed.

There are no secrets. When we ask writers about their process or scientists about their methods or inventors where they get their ideas from, we are hoping for something that doesn't exist: a trick, recipe, or ritual to summon the magic—an alternative to work. There isn't one. To create is to work. It is that easy and that hard.

With the myth gone, we have a choice. If we can create without genius or epiphany, then the only thing stopping us from creating is us. There is an arsenal of ways to say no to creating. One, *it is not easy*, has already been addressed. It is not easy. It is work.

Another is *I have no time*. But time is the great equalizer, the same for all: twenty-four hours every day, seven days every week, every life a length unknown, for richest and poorest and all between. We mean *we have no spare time,* a blunt blade in a world whose bestselling literary series was begun by a single mother writing in Edinburgh's cafés when her infant daughter slept, where a career more than fifty novels long was started by a laundry worker in the furnace room of a trailer in Maine, where world-changing philosophy was composed in a Parisian jail by a prisoner awaiting the guillotine, and where three centuries of physics were overturned in a year by a man with a permanent position as a patent examiner. There is time.

The third no is the big one, the gun to the head of our dreams. Its endless variations all say the same thing: *I can't.* Here is the sour fruit of the myth that only the special can create. None of us think we are special, not in the middle of the night, when our faces fluoresce in the bathroom mirror. *I can't,* we say. *I can't because I am not special.*

We *are* special, but that does not matter right now. What matters is that we do not have to be. The creativity myth is a mistake born of a need to explain extraordinary outcomes with extraordinary acts and extraordinary characters, a misunderstanding of the truth that creation comes from ordinary people and ordinary work. Special is not necessary.

All that is necessary is to begin. *I can't* is not true once we begin. Our first creative step is unlikely to be good. Imagination needs iteration. New things do not flow finished into the world. Ideas that seem powerful in the privacy of our head teeter weakly when we set them on our desk. But every beginning is beautiful. The virtue of a first sketch is that it breaks the blank page. It is the spark of life in the swamp. Its quality is not important. The only bad draft is the one we do not write.

How to create? Why create? The rest of the book is about how and why. What to create? Only you can decide that. You may know. You may have an idea like an itch. But if you do not, don't worry. How and what are connected: one leads to the other.

CHAPTER 2

THINKING IS LIKE WALKING

1 | KARL

Berlin once stood at the center of the creative world. The city's theaters reverberated with debuts by Max Reinhardt and Bertolt Brecht. Its nightclubs hosted bawdy burlesque *Kabarett*. Albert Einstein ascended its Academy of Sciences. Thomas Mann prophesied the perils of National Socialism. The movies *Metropolis* and *Nosferatu* premiered to packed houses. Berliners called it the Golden Age: the years of Marlene Dietrich, Greta Garbo, Joseph Pilates, Rudolf Steiner, and Fritz Lang.

It was a time and place for thinking about thinking. In Berlin, German psychologists were having radical thoughts about how the human mind works. Otto Selz, a professor in Mannheim, far to the southwest, sowed the seed: he was among the first to propose that thinking was a process that could be scrutinized and described. For most of his contemporaries, the mind was magic and mystery. For Selz, it was mechanism.

But as the 1930s began, Otto Selz heard the boots of doom

approaching. He was Jewish. Hitler was rising. Berlin's celebration of creation was turning apocalyptic. Destruction was coming.

Selz had been asking psychological questions: How did a mind work? Could he measure it? What could he prove? Now he was also asking practical ones: What was going to happen to him? Could he escape it? How much time remained?

And—equally important to him—would his thoughts survive if he did not? His chance to pass them on was brief. In 1933, the Nazis prevented him from working and prohibited others from citing him. His name disappeared from the literature.

But at least one Berliner knew Selz's work. Karl Duncker was thirty years old when the Nazis banned Otto Selz. Duncker was not Jewish. His appearance was Aryan: fair skin, flaxen hair, and faceted jaw. He was no safer for it. His ex-wife was Jewish, and his parents were Communists. He made two applications to become a professor at the University of Berlin. Both were rejected despite his excellent academic record. In 1935, the school fired him from his job as a researcher. He published his masterwork, *On Problem Solving*—in which he defied the Nazis by citing Selz ten times—and fled to the United States.

The Golden Age was over. Novelist Christopher Isherwood, teaching English in Berlin, captured its passing:

> Today the sun is brilliantly shining; it is quite mild and warm. I
> go out for my last morning walk, without an overcoat or hat. The
> sun shines, and Hitler is master of this city. The sun shines, and
> dozens of my friends—my pupils at the Workers' School, the men
> and women I met are in prison, possibly dead or being tortured
> to death. I catch sight of my face in the mirror of a shop, and am
> horrified to see that I am smiling. You can't help smiling, in such
> beautiful weather. The trams are going up and down the Kleist-
> strasse, just as usual. They, and the people on the pavement, and
> the tea-cosy dome of the Nollendorfplatz station have an air of

curious familiarity, of striking resemblance to something one remembers as normal and pleasant in the past.

Duncker took a position in the psychology department of Swarthmore College, in Pennsylvania. In 1939, he produced his first paper since arriving in America, coauthored with Isadore Krechevsky, an immigrant who'd left the tiny Lithuanian village of Sventijánskas as a young boy to escape Russian anti-Semitism. Krechevsky, whose encounters with prejudice in the United States had brought him to the edge of abandoning his academic career, was the first American Duncker inspired.

The joint paper, "On Solution-Achievement," published in *Psychological Review,* marks the moment in the history of the mind when America met Berlin. Krechevsky, in the American style of the time, studied learning in rats. Duncker studied thinking in humans. This was so unusual that Duncker had to clarify what thinking meant: "The functional sense of problem-solving, not a special, e.g., imageless, kind of representation."

In the paper, the two men agreed that solving problems required "a number of intermediate steps," but Krechevsky noted a crucial difference between Duncker's ideas and the ones that were prevalent in America: "There is in Duncker's analysis one major concept which does not find a close parallel in American psychology: in his experiments, the solution to the problem is a meaningful one. The organism can bring to bear experiences from other occasions and comparatively few general experiences can be utilized for problem-solving."

Duncker had made his first mark. American psychologists experimented on animals and spoke of *organisms:* train-your-rat psychology. Duncker cared about human minds and meaningful problems. He put his foot to a shovel and broke ground for a cognitive revolution that would take twenty years to build.

In Germany, the Nazis arrested Otto Selz and took him to Dachau, their first concentration camp. They held him there for five weeks.

Duncker published his second paper, on the relationship between familiarity and perception, in the *American Journal of Psychology*.

In Russia, his brother Wolfgang was captured in Stalin's Great Purge and murdered in the gulag.

Duncker's third paper of the year was published in the pioneering journal of philosophy and psychology *Mind*. His subject was the psychology of ethics. Duncker wanted to understand why people's moral values varied so much. The paper was nuanced, comprehensive, and poignant. A man devoted to discovering how humans think was trying to make sense of the end of Berlin:

> The motive "for the benefit of the State" depends upon whether the State is felt to be the embodiment of the highest values of life or merely a sort of police-station. On the whole moral judgments are based upon the standard meanings of the society in question. Its chief aim is not to be "just," but to instigate and to enforce its standard meanings and conducts. It is this function which interferes with a purely ethical conduct.

Here was his answer. States can replace ethics with edicts.
At the end of February 1940, Karl Duncker wrote something else.

Dear Mother,

You have been good to me.
*   Don't condemn me.*

He drove to nearby Fullerton and, while sitting in his car, shot himself in the head with a pistol. He was thirty-seven.

In Amsterdam, the Nazis captured Otto Selz, took him to Auschwitz, and murdered him.

In Berkeley, the University of California awarded a professorship in psychology to a man named David Krech. He had changed his name from Isadore Krechevsky. He was Duncker's first American coauthor. He went on to have a storied thirty-year career specializing in the mechanics of memory and stimulation.

Krech was one of many people Duncker influenced. Duncker carried Germany's best and most radical ideas about thinking to the United States and started a revolution he did not live to see. He was a message in a bottle cast from the shores of a dying Berlin. The bottle broke, but not before it had delivered its message.

2 | THE QUESTION OF FINDING

Duncker's monograph *On Problem Solving*, which he published in 1935 as he was fleeing Germany, led to a transformation in the science of brain and mind known as the "cognitive revolution" that laid the foundation for our understanding of how people create. For many reasons, including its references to Otto Selz, *On Problem Solving* was verboten in Hitler's nation. War arrived. Berlin burned. Copies became rare.

Then, five years after Duncker's suicide, one of his former students, Lynne Lees, revived the monograph by translating it into English and presenting his bold agenda—"to study productive thinking"—to the world.

Duncker rejected studies of great thinkers. He likened them to lightning—a dramatic display of something "better investigated in little sparks within the laboratory." He used "practical and mathematical problems because such material is more suitable for experimentation," but he made it clear that he was studying *thought*, not puzzles or math.

It did not matter what someone was thinking *about* — the "essential features of problem-solving are independent of the thought-material."

For millennia, people had been herded into categories: civilized and savage, Caucasian and Negro, man and woman, Gentile and Jew, rich and poor, capitalist and Communist, genius and dullard, gifted and non-gifted. Category determined capacity. By the 1940s, these divisions had been reinforced by "scientists" who evoked the innate potential to organize the human race like a zoo and lock "different" people into cages, sometimes literal ones. Then a Gentile who married a Jew, a son of Communists who emigrated to live among capitalists, a man who collaborated with Jews and women and who had witnessed the horrors caused by the fraud of measuring humanity, showed that human thought has an essence unaffected by scale, subject, or thinker — that our minds all work the same way.

It was radical and controversial, and it shifted the shape of psychology. Duncker's approach was simple. He gave people problems and asked them to think aloud as they tried to solve them. In this way, he saw the structure of thought.

Thinking is finding a way to achieve a goal that cannot be attained by an obvious action. We want to accomplish something but do not know how, so before we can act we must think. But *how* do we think? Or, as Duncker phrased it, what is the answer to "the specific question of finding: In what way can a meaningful solution be found?"

We all use the same process for thinking, just as we all use the same process for walking. It is the same whether the problem is big or small, whether the solution is something new or something logical, whether the thinker is a Nobel laureate or a child. There is no "creative thinking," just as there is no "creative walking." Creation is a *result* — a place thinking may lead us. Before we can know how to create, we must know how to think.

Duncker deployed an array of experiments. They included the

Abcabc Problem, which asked high school students to work out why numbers in the form 123,123 and 234,234 are always divisible by 13; the Stick Problem, in which babies as young as eight months old were given a stick that enabled them to reach a remote toy; the Cork Problem, where a piece of wood had to be inserted into a door frame even though it was not as long as the door was wide; and the Box Problem, where candles had to be attached to a wall by selecting from objects including thumb tacks and various boxes. Duncker varied his experiments many times until he understood how people think, what helps, and what gets in the way.

One of his conclusions: "If a situation is introduced in a certain perceptual structure, thinking achieves a contrary structure only against the resistance of the former structure."

Or: old ideas obstruct new ones.

And this was the case with Duncker's work. Few psychologists read or understood *On Problem Solving* in its entirety—not because it was complicated but because old ideas made them resist it. Today the monograph is known mostly for the Box Problem, which has been given the misnomer the Candle Problem and also redesigned. It attracted more attention than all the others. Psychologists and people who write about creation have been discussing it for more than fifty years. Here is its modern incarnation:

Picture yourself in a room with a wooden door. The room contains a candle, a book of matches, and a box of tacks. Using only these things, how would you attach the candle to the door so that you can light it, have it burn normally, and create light to read by?

People usually think of three solutions. One is to melt part of the candle and use the melted wax to fix the candle to the door. Another is to tack the candle to the door. Both work, but not very well. The third solution, which occurs to only a minority of people, is to empty out the tack box, tack *that* to the wall, and use it to hold the candle.

This last solution has a feature the others do not: one of the items, the box, is used for something other than its original purpose. At some point the person solving the problem stops seeing it as a thing for holding the tacks and starts seeing it as a thing for holding the candle.

This shift, sometimes called an *insight,* is considered important by some people who think about creating. They suspect that there is something remarkable about seeing the box differently, that the shift is a leap like the one we experience when we look at that picture of a vase that might be two faces or the old lady that could be a young lady or the duck that may be a rabbit. Once we make this "leap," the problem is solved.

Following Duncker's lead, psychologists have created many similar puzzles. Examples include the Charlie Problem:

Dan comes home one night after work, as usual. He opens the door and steps into the living room. On the floor he sees Charlie lying dead. There is water on the floor, as well as some pieces of glass. Tom is also in the room. Dan takes one quick glance at the scene and immediately knows what happened. How did Charlie die?

And the Prisoner and Rope Problem:

A prisoner was attempting to escape from a tower. He found in his cell a rope that was half long enough to permit him to reach the ground safely. He divided the rope in half, tied the two parts together, and escaped. How could he have done this?

And the Nine-Dot Problem:

Picture three rows of three dots, evenly spaced to resemble a square. Join the dots using only four straight lines without taking your pencil or pen off the paper.

All are solved the same way: by the equivalent of realizing that the faces are also a vase. Charlie is not a person but a fish. Tom is not a person but a cat. Tom knocked over Charlie's fish bowl, and Charlie died. The prisoner did not "divide" the rope in half widthwise, as we naturally imagine, but lengthwise. The nine dots are joined by drawing lines that extend beyond the "square" created by the dots. This is the source of the cliché "thinking outside the box."

Does this mean minds leap? We can answer that question with one more problem, the Speckled Band:

Julia sleeps in a locked room. Beside her bed is a bell pull for summoning the housekeeper. Above the bell pull is a ventilator that connects to the room next door. That room contains a safe, a dog leash, and a saucer of milk. One night Julia screams. There is a whistle and a clang. She is found dying with a burnt match in her hand. There are no signs of violence. There are no pets in the house. Her room had remained locked. Her last words are "the speckled band." How did Julia die?

This is not a psychology problem. It is a summary of a Sherlock Holmes story written by Arthur Conan Doyle in 1892. Julia died from the bite of a poisonous snake trained to crawl through the ventilator and down the bell pull, then return when her murderer whistled. He kept the snake on the leash and fed it the milk. The clang is the sound of him hiding the snake in the safe after the murder. Upon being bitten, Julia lit a match for illumination and glimpsed the snake, which looked to her like a "speckled band."

Holmes works this out by observing that the only way into the locked room is through the ventilator. He deduces that because Julia died quickly and without obvious signs of violence, she was probably poisoned. Something small and poisonous therefore passed through the ventilator. The dog leash suggests an animal, rather than a gas, and the saucer of milk rules out an insect such as a spider. Julia's dying words about a speckled band, which initially seem cryptic, now sound like a description of the most likely remaining solution: a snake, trained to respond to its master's whistle. The clang shows that the snake is in the safe.

Holmes is a fictional character famous for detection, not creation.

He describes his process as "observation and deduction: eliminate all other factors, and the one which remains must be the truth." He does not solve Julia's murder with a creative leap. The "insight" that begins his process of deduction—that the only way into the locked room is via the ventilator—is an observation. The surprising solution that a snake killed her follows.

Minds do not leap. Observation, evaluation, and iteration, not sudden shifts of perception, solve problems and lead us to creation. We can see this using Duncker's technique: observing people solving his most famous problem.

3 | STEPS, NOT LEAPS

Many people do not think using words, but we can all verbalize our thoughts without affecting our problem-solving skills. Listening to the mind shows how thinking works. Robert Weisberg asked people to think aloud as they worked on Duncker's Box Problem. He changed the problem by including nails as well as tacks and substituting a piece of cardboard for the wooden door. The people he worked with had the objects in front of them. They were asked to imagine solutions but not build them.

Here are the thoughts of three people who did *not* think of using the tack box as a candleholder:

PERSON 1: "Melt the candle and try to stick it up. Candle coming out vertically on a nail, but it will break. Put the candle sideways and nail it up. The candle looks heavy. Put a nail or two nails in the side of the candle, but it might not stay up. I could . . . no, I couldn't do that."

PERSON 2: "I'm looking at the nails, but they won't penetrate but otherwise how will the candle stick? Put a nail through the vertical candle. Put a nail through the candle held horizontal. Can't use the matches. Put nails in the wick and under the candle . . ."

PERSON 3: "I was thinking you could take a nail and bang it through, but that would split the candle, so use the matches to melt enough wax, then use the nails—no good. Bang the nails in close together and put the candle on them. . . ."

And here are the thoughts of three people who *did* think of using the tack box to hold the candle:

PERSON 4: "Candle has to burn straight, so if I took a nail and put it through the candle and cardboard . . . [10 second pause] . . . if I took several nails and made a row and set the candle on that. If I took the nails out of the box, nailed the box to the wall."

PERSON 5: "Melt wax and use it to stick the candle up. Take a nail—the nail won't go through the candle. Put nails around the candle or under the candle to hold it. Put the candle in the nail box—it wouldn't work, the box would rip."

PERSON 6: "Light a match and see if I could get wax up on the cardboard. Push a nail through the candle into the cardboard. I'm looking at the matches to see if the idea would work. I'm trying to get more combinations with the nails. Build a base for the candle with the nails like a rectangle. Better yet, use the box. Put two nails into the cardboard, put the box on them, melt some wax and put the candle into the box with the wax and it'll stand."

This is how we think. *Everyone* who thinks of using the tack box gets there the same way. After eliminating other ideas, they think of building a platform out of nails, then think of using the tack box as the platform. There is no sudden shift of perception. We move from known to new in small steps. In every case, the pattern is the same: begin with something familiar, evaluate it, solve any problems, and repeat until a satisfactory solution is found. Duncker discovered this in the 1930s:

"Successful people arrived at the solution in this way: they started from tacks and looked for a 'platform to be fastened to the door with tacks.'"

Evaluation directs iteration. Person 3 decides to "bang the nails in close together and put the candle on them" and evaluates this as satisfactory. Person 4 evaluates this as *un*satisfactory so takes one more step: use the tack box. Person 5 also takes this step, the solution Duncker sought for his problem, but makes the opposite evaluation: it won't work. Person 6 takes the most steps of all and, as a result, improves Duncker's solution by using melted wax to stabilize the candle.

Creating is taking steps, not making leaps: find a problem, solve it, and repeat. Most steps wins. The best artists, scientists, engineers, inventors, entrepreneurs, and other creators are the ones who keep taking steps by finding new problems, new solutions, and then new problems again. The root of innovation is exactly the same as it was when our species was born: looking at something and thinking, "I can make this better."

Six undergraduates talking their way through a puzzle is not enough for generalization, nor is twenty-five, which is how many Weisberg asked to talk out loud, nor even 376, which is how many tried the Box Problem in his experiment. But these results do undermine a vital premise of the creativity myth: that creating requires leaps of extraordinary thinking. It does not. Ordinary thinking works.

4 | AHA!

There is an alternative to the theory that creation comes from ordinary thinking: the idea, proposed by psychologists Pamela Auble,

Jeffrey Franks, and Salvatore Soraci, writer Jonah Lehrer, and many others, that many of the best creations come from an extraordinary moment of sudden inspiration, sometimes called the "eureka effect" or "aha! moment." Ideas start as caterpillars of the conscious mind, become cocoons in the unconscious, then fly out like butterflies. This moment results in excitement and possibly exclamation. The key to creating is to cultivate more of these moments.

People who believe this will have many reasonable objections to the proposal that creation comes from ordinary thinking. There are documented cases of great creators having aha! moments. Many people have experienced frustration with a problem and set it aside only to have the solution come to them. Neurologists seeking the source of such moments are discovering interesting things. The aha! moment is woven into our world. Oprah Winfrey has trademarked it. How can ordinary thinking explain this?

The most frequently cited story of an aha! moment was first made famous by a Roman architect named Vitruvius.

Vitruvius says that when the Greek general Hiero was crowned king of Syracuse, in Sicily, twenty-three hundred years ago, he celebrated by giving a craftsman some gold and asking him to create a golden wreath. The craftsman duly delivered a wreath with the same weight as the gold that Hiero had provided, but Hiero suspected that he had been tricked and that most of the wreath was made of silver. Hiero asked Syracuse's greatest thinker, a twenty-two-year-old named Archimedes, to find the truth: was the wreath pure gold or a mixture of gold and silver? According to Vitruvius, Archimedes then took a bath. The lower he sank, the more the water overflowed. This gave him an idea. Archimedes ran home naked, shouting, *"Eureka, eureka!"*—"I have found it, I have found it!" He made two objects that were both the same weight as the wreath, one in gold and one in silver, and submerged each of them in water and measured how much water overflowed. The silver object displaced more water than the gold object. Then Archimedes submerged Hiero's "golden" wreath into the water. It displaced more water than the same weight of pure gold, proving that it had been adulterated with silver or some other substance.

This story about Archimedes, which Vitruvius told two centuries after the fact, is almost certainly not true. The method Vitruvius describes does not work, as Archimedes would have known. Galileo pointed this out in a paper called "La Bilancetta" ("The Little Balance"), which calls the method of comparing gold and silver Vitruvius describes "altogether false." The tiny differences in the amount of water displaced by the gold, the silver, and the wreath would have been too hard to measure. Surface tension and drops of water that remained on the wreath would have caused other problems. Galileo's paper shows the method Archimedes probably used, based on Archimedes's own work: weighing the wreath underwater. Buoyancy, not displacement, is the key to solving the problem. Overflowing a bath is unlikely to have inspired this.

But let's take Vitruvius's story at face value. He says that Archimedes, "while the case was still on his mind, happened to go to the bath, and on getting into a tub observed that the more his body sank into it, the more water ran out over the tub. As this pointed out the way to explain the case in question, without a moment's delay, and transported with joy, he jumped out of the tub and rushed home naked, crying with a loud voice that he had found what he was seeking; for as he ran he shouted repeatedly in Greek, 'Eureka, eureka.'"

Or: Archimedes's eureka moment came from an observation he made *while thinking about the problem*. At best, the bath is like the platform of nails in the Weisberg experiments: it is the one thing that leads to another. If it happened at all, Archimedes's legendary shout of *"Eureka"* did not come from an aha! moment but from the simple joy of solving a problem with ordinary thinking.

Another famous example of an aha! moment comes from Samuel

Taylor Coleridge, who claimed his poem "Kubla Khan" was written in a dream. According to Coleridge's preface:

> In the summer of the year 1797, the Author, then in ill health, had retired to a lonely farmhouse. An anodyne had been prescribed, from the effects of which he fell asleep in his chair at the moment that he was reading, "Here the Khan Kubla commanded a palace to be built, and a stately garden thereunto. And thus ten miles of fertile ground were inclosed with a wall." The author continued for about three hours in a profound sleep during which time he could not have composed less than from two to three hundred lines without any sensation or consciousness of effort. On awaking he eagerly wrote down the lines; at this moment he was unfortunately called out by a person on business from Porlock and on his return to his room, found all the rest had passed away.

This gave the poem—subtitled "A Vision in a Dream"—an aura of mystery and romance that continues to this day. But Coleridge is misleading us. The anodyne, or painkiller, which he says he had been prescribed was opium dissolved in alcohol—a substance to which Coleridge was addicted. A trance of three to four hours is a classic opium-induced state, which can be euphoric and hallucinogenic. Coleridge's movements in the summer of 1797 are well known. He had no time to retire to a lonely farmhouse. The person from Porlock may have been fictitious and an excuse for not finishing the poem. Coleridge used a similar device—a fake letter from a friend—to excuse the incompleteness of another work, his *Biographia Literaria.* The preface claims the poem was composed during sleep, then written automatically. But in 1934, an earlier manuscript of "Kubla Khan" was found that differs from the published poem. Among many changes, "From forth this Chasm with hideous turmoil seething" became "*And from* this chasm, with *ceaseless* turmoil seething"; "So twice six miles of fertile ground /

With Walls and Towers were compass'd round" was changed to "So twice *five* miles of fertile ground / With walls and towers were *girdled* round"; "mount Amora" was rewritten as "mount *Amara*"—a reference to Milton's *Paradise Lost*—then, finally, "mount *Abora.*" The origin story changed, too. Coleridge says the poem was "composed in a sort of reverie brought on by two grains of opium" in the fall, rather than appearing complete during a sleep in the summer.

These are minor changes, but they show conscious thought, not unconscious automation. "Kubla Khan" may or may not have started in a dream, but ordinary thinking finished it.

A third frequently told story about an aha! moment comes from 1865, when chemist August Kekulé discovered the ringlike structure of benzene. Twenty-five years after making this discovery, Kekulé said, in a speech to the German Chemical Society:

> I was sitting writing at my textbook but the work did not progress; my thoughts were elsewhere. I turned my chair to the fire and dozed. Again the atoms were gamboling before my eyes. This time the smaller groups kept modestly in the background. My mental eye, rendered more acute by repeated visions of the kind, could now distinguish larger structures of manifold conformation: long rows, sometimes more closely fitted together all twining and twisting in snake-like motion. But look! What was that? One of the snakes had seized hold of its own tail, and the form whirled mockingly before my eyes. As if by a flash of lightning I awoke; and this time also I spent the rest of the night in working out the consequences of the hypothesis.

Robert Weisberg points out that the word Kekulé used was *halbschlaf,* or "half-sleep," which is often translated as "reverie." Kekulé was not sleeping. He was daydreaming. His dream is often described as a vision of a snake biting its tail. But Kekulé says he saw *atoms* twisting in a *snake-like motion*. When he later describes one of the snakes seizing its tail, he is referring back to his analogy. He is not seeing a snake. This is a case of visual imagination helping solve a problem, not an aha! moment happening in a dream.

A sudden revelation has also been attributed to Einstein, who was stuck for a year while developing the special theory of relativity and went to a friend for help. "It was a beautiful day when I visited him with this problem," he said. "I started the conversation with him in the following way: 'Recently I have been working on a difficult problem. Today I come here to battle against that problem with you.' We discussed every aspect of this problem. Then suddenly I understood where the key to this problem lay. Next day, I came back to him again and said to him, without even saying hello, 'Thank you. I've completely solved the problem.'"

Was this a flash of inspiration? No. In Einstein's own words: "I was led to it by steps." All stories of aha! moments—and there are surprisingly few—are like these: anecdotal, often apocryphal, and unable to survive scrutiny.

And there has been a lot of scrutiny: in the last few decades of the twentieth century, many psychologists believed that creation comes from a period of unconscious thinking they called "incubation," followed by an emotion they called "the feeling of knowing," followed by an aha! moment, or "insight." These psychologists conducted hundreds of experiments designed to validate their hypothesis.

For example, in 1982, two researchers at the University of Colorado tested the feeling of knowing with thirty people in an experiment lasting nineteen days. They showed the subjects pictures of entertainers and asked them to recall the entertainers' names. Only 4 percent of memories were recovered spontaneously, most of them by the same four people. All the other memories were recovered by ordinary thinking: gradually working through the problem by remembering, for example, that the entertainer was a movie star in the 1950s, that he had appeared in an Alfred Hitchcock movie where he was chased by a crop duster, that the movie was called *North by Northwest,* and, finally, that his name was Cary Grant. The study's conclusion? Even the "spontaneous" memories had probably also come from ordinary thinking, and there was no support for unconscious mental processing as a way of recovering memories. Other studies into the feeling of knowing have had similar results.

And what of incubation? An academic named Robert Olton spent many years at the University of California, Berkeley, trying to prove that incubation exists. In one experiment he sorted 160 people into ten groups and asked them to solve an insight problem called the Farm Problem, which involves dividing an L-shaped "farm" into four parts of the same size and shape. The solution is novel—you have to make four smaller L shapes in various orientations. Every subject was tested individually and given thirty minutes to solve the problem. To see if taking a break from thinking—that is, incubating—made a difference, some subjects were given a fifteen-minute break. During this break, some people could do whatever they wanted; others were given mental work like counting backward in threes, or were asked to talk about the problem out loud, or were told to relax in a room with a comfortable chair, dim lights, and soft music. Each activity tested a different idea about how incubation works.

But the results were the same for every group. People who worked continuously performed as well as people given a period for incubation. People given a period for incubation performed as well as one another, regardless of what they did during incubation. Olton sliced the data many ways, looking for evidence that incubation worked, but was forced to conclude, "The major finding of this study is that no evidence of incubation was apparent under any condition, even under those where its appearance would seem most likely." He called this "an inexorably negative finding." He was also unable to replicate any of the positive results others had reported. "To our knowledge," he wrote, "no study reporting evidence of incubation in problem solving has survived replication by an independent investigator."

Olton suggested that one explanation for the lack of evidence supporting incubation was flawed experiments. But he added, "A second, more radical, explanation of our results is to accept them at face value and to question the existence of incubation as an objectively demonstrable phenomenon. That is, incubation may be something of an illusion, perhaps rendered impressive by selective recall of the few but vivid occasions on which great progress followed separation from a problem and forgetting of the many occasions when it did not."

To his credit, Robert Olton did not give up. He designed a different study, this time using experts trying to solve a problem in their area of expertise—chess players and a chess problem—in the hope that this would give better results than undergraduates with an insight problem. Half his subjects worked continuously and half were given a break, during which they were asked not to think about the problem. Again, the break made no difference. Both groups performed equally well. Olton, initially a believer in incubation, was forced to doubt its existence. His despair was evident in the subtitle of the paper he wrote about the study: "Searching for the Elusive." The paper concluded, "We simply didn't find incubation."

Most researchers now regard incubation as folk psychology—a popular belief but wrong. Almost all of the evidence suggests the same thing: Caterpillars do not cocoon in the unconscious mind. The butterflies of creation come from conscious thinking.

5 | THE SECRET OF STEVE

Karl Duncker wrote that the act of creation starts with one of two questions: "'Why doesn't it work?' or, 'What should I change to make it work?'"

These sound simple, but answering them can lead to extraordinary results. One of the best examples comes from Steve Jobs, cofounder and CEO of Apple Inc. When Jobs announced Apple's first cell phone, the iPhone, in 2007, he said:

> The most advanced phones are called smartphones. They are definitely a little smarter, but they actually are harder to use. They all have these keyboards that are there whether you need them or not. How do you solve this? We solved it in computers 20 years ago. We solved it with a screen that could display anything. What we're going to do is get rid of all these buttons and just make a giant screen. We don't want to carry around a mouse. We're going to use a stylus. No. You have to get them and put them away, and you lose them. We're going to use our fingers.

It is no coincidence that Jobs sounds like one of Duncker's subjects thinking aloud while trying to attach a candle to a door. The step-by-step process is the same. Problem: Smarter phones are harder to use because they have permanent keyboards. Solution: A big screen and a pointer. Problem: What kind of pointer? Solution: A mouse. Problem: We don't want to carry a mouse around. Solution: A stylus. Problem: A stylus might get lost. Solution: Use our fingers.

Apple sold 4 million phones in 2007, 14 million in 2008, 29 million in 2009, 40 million in 2010, and 82 million in 2011, for a total of 169 million sold in its first five years in the phone business, despite charging a higher price than its competitors did. How?

For several years, starting around 2002, I was a member of the research advisory board of a company that made cell phones. Every year it gave me its latest phone. I found each one harder to use than the last, as did other board members. It was no secret that Apple might enter the cell phone market, but the risk was always dismissed, since Apple had never made a phone. A few months after Apple's phone became available, the board met and I asked what the company thought of it. The chief engineer said, "It has a really bad microphone."

This was true, irrelevant, and revealing. This company thought smartphones were phones, only smarter. They had made some of the first cell phones, which, of course, had buttons on them. These had been successful. So, as they added smarts, they added buttons. They thought a good phone provided a good phone call and the smart stuff was a bonus.

Apple made computers. For Apple, as Jobs's announcement made clear, a smartphone was not a phone. It was a computer for your pocket that, among other things, made calls. Making computers was a problem that Apple, as Jobs described it, had "solved" twenty years ago. It did not matter that Apple had never made a phone. It did matter that phone makers had never made a computer. The company I was advising, once a leading phone manufacturer, lost a large amount of money in 2007, saw its market share collapse, and was eventually sold.

"Why doesn't it work?" deceives us with its simplicity. The first challenge is to ask it. The chief engineer did not ask this question about his phones. He saw rising sales and happy customers and so assumed that nothing was broken and there was nothing to fix.

But Sales + Customers = Nothing Broken is a formula for corporate cyanide. Most big companies that die kill themselves drinking it. Complacency is an enemy. "If it ain't broke, don't fix it" is an impossible idiom. No matter the sales and customer satisfaction, there is always something to fix. Asking, "Why doesn't it work?" is creation inhaling. Answering is creation breathing out. Innovation suffocates without it.

"Why doesn't it work?" has the pull of a polestar. It sets creation's direction. For Jobs and the iPhone, the critical point of departure was not finding a solution but seeing a problem: the problem of keyboards making smarter phones harder to use. Everything else followed.

Apple was not unique. Korean electronics giant LG launched a product much like the iPhone before the iPhone was announced.

The LG Prada had a full-sized touch screen, won design awards, and sold a million units. When Apple's very similar direction—a big touch screen—was revealed, competitors built near replicas within months. These other companies could make an iPhone, but they could not conceive one. They could not look at their existing products and ask, "Why doesn't it work?"

The secret of Steve was evident in 1983, during the sunrise of the personal computer, when he spoke at a design conference in Aspen, Colorado. There was no stage, and there were no visual aids. Jobs stood behind a lectern with yearbook hair, a thin white shirt, its sleeves folded as far as his forearms, and—"they paid me sixty dollars, so I wore a tie"—a pink-and-green bow tie. The audience was small. He gestured widely as he envisioned "portable computers with radio links," "electronic mailboxes," and "electronic maps." Apple Computer, of which Jobs was then cofounder and a director, was a six-year-old start-up playing David to IBM's Goliath. Apple's sling was sales; it had sold more personal computers than any other company in 1981 and 1982. But despite his optimism, Jobs was dissatisfied:

> If you look at computers, they look like garbage. All the great
> product designers are off designing automobiles or buildings but
> hardly any of them are designing computers. We're going to sell
> ten million computers in 1986. Whether they look like a piece a
> shit or they look great. There are going to be these new objects in
> everyone's working environment, in everyone's educational envi-
> ronment, in everyone's home environment. And we have a shot
> at putting a great object there. Or if we don't, we're going to put
> one more piece of junk there. By 1986 or 1987 people are going to
> be spending more time interacting with these machines than they
> spend in a car. And so industrial design, software design, and how
> people interact with these things must be given the consideration
> that we give automobiles today, if not a lot more.

Twenty-eight years later, Walt Mossberg, technology columnist for the *Wall Street Journal,* described a similar discussion that happened near the end of Jobs's life: "One minute he'd be talking about sweeping ideas for the digital revolution. The next about why Apple's current products were awful, and how a color, or angle, or curve, or icon was embarrassing."

A good salesman sells everybody. A great salesman sells everybody but himself. What made Steve Jobs think differently was not genius, passion, or vision. It was his refusal to believe that sales and customers meant nothing was broken. He enshrined this in the name of the street encircling Apple's campus: Infinite Loop. The secret of Steve was that he was never satisfied. He devoted his life to asking, "Why doesn't it work?" and "What should I change to make it work?"

But hang on. Surely there is an alternative to starting by asking, "Why doesn't it work?" What if you simply start with a good idea?

Ideas are a staple of myths about creating; they even have their own symbol, the lightbulb. That comes from 1919, the age of silent movies, a decade before Mickey Mouse, when the world's favorite animated animal was Felix the Cat. Felix was black, white, and mischievous. Symbols and numbers would appear above his head, and sometimes he would grab them to use as props. Question marks became ladders, musical notes became vehicles, exclamation points became baseball bats, and the number 3 became horns he used to turn the tables on a bull. One symbol lived long after the cat: when Felix had an idea, a lightbulb appeared above his head. Lightbulbs have represented ideas ever since. Psychologists adopted the image: after 1926, they often called having an idea *illumination*.

The creativity myth confuses having ideas with the actual work of creating. Books with titles like *Making Ideas Happen, How to Get Ideas, The Idea Hunter,* and *IdeaSpotting* emphasize idea generation, and idea-generation techniques abound. The most famous is brainstorming, invented by advertising executive Alex Osborn in 1939 and first published in 1942 in his book *How to Think Up.* This is a typical description, from James Manktelow, founder and CEO of MindTools, a company that promotes brainstorming as a way to "develop creative solutions to business problems":

> Brainstorming is often used in a business setting to encourage teams to come up with original ideas. It's a freewheeling meeting format, in which the leader sets out the problem that needs to be solved. Participants then suggest ideas for solving the problem, and build on ideas suggested by others. A firm rule is that ideas must not be criticized—they can be completely wacky and way out. This frees people up to explore ideas creatively and break out of established thinking patterns. As well as generating some great solutions to specific problems, brainstorming can be a lot of fun.

Osborn claimed significant success for his technique. As one example of brainstorming's effectiveness, he cited a group of United States Treasury employees who came up with 103 ideas for selling savings bonds in forty minutes. Corporations and institutions including DuPont, IBM, and the United States government soon adopted brainstorming. By the end of the twentieth century, its origins forgotten, brainstorming had become a reflex approach to creating in many organizations and had entered the jargon of business as both a noun and a verb. It is now so common that few people question it. Everybody brainstorms; therefore, brainstorming is good.

But does it work?

Claims about the success of brainstorming rest on easily tested assumptions. One assumption is that groups produce more ideas than individuals. Researchers in Minnesota tested this with scientists and advertising executives from the 3M Company. Half the subjects worked in groups of four. The other half worked alone, and then their results were randomly combined as if they had worked in a group, with duplicate ideas counted only once. In every case, four people working individually generated between 30 to 40 percent more ideas than four people working in a group. Their results were of a higher quality, too: independent judges assessed the work and found that the individuals produced better ideas than the groups.

Follow-up research tested whether larger groups performed any better. In one study, 168 people were either divided into teams of five, seven, or nine or asked to work individually. The research confirmed that working individually is more productive than working in groups. It also showed that productivity decreases as group size increases. The conclusion: "Group brainstorming, over a wide range of group sizes, inhibits rather than facilitates creative thinking." The groups produced fewer and worse results because they were more likely to get fixated on one idea and because, despite all exhortations to the contrary, some members felt inhibited and refrained from full participation.

Another assumption of brainstorming is that suspending judgment is better than assessing ideas as they appear. Researchers in Indiana tested this by asking groups of students to think of brand names for three different products. Half of the groups were told to refrain from criticism and half were told to criticize as they went along. Once again, independent judges assessed the quality of each idea. The groups that did not stop to criticize produced more ideas, but both groups produced the same number of good ideas. Deferring criticism added only bad ideas. Subsequent studies have reinforced this.

Research into brainstorming has a clear conclusion. The best way to create is to work alone and evaluate solutions as they occur. The worst way to create is to work in large groups and defer criticism. Steve Wozniak, Steve Jobs's cofounder at Apple and the inventor of its first computer, offers the same advice: "Work alone. You're going to be best able to design revolutionary products and features if you're working on your own. Not on a committee. Not on a team."

Brainstorming fails because it is an explicit rejection of ordinary thinking—all leaps and no steps—and because of its unstated assumption that having ideas is the same as creating. Partly as a result, almost everybody has the idea that ideas are important. According to novelist Stephen King, the question authors signing books get asked most often—and are least able to answer—is "Where do you get your ideas from?"

Ideas are like seeds: they are abundant, and most of them never grow into anything. Also, ideas are seldom original. Ask several independent groups to brainstorm on the same topic at the same time, and you will likely get many of the same ideas. This is not a limitation of brainstorming; it is true of all creation. Because everything arises from steps, not leaps, most things are invented in several places simultaneously when different people walk the same path, each unaware of the others. For example, four different people discovered sunspots independently in 1611; five people invented the steamboat between 1802 and 1807; six people conceived of the electric railroad between 1835 and 1850; and two people invented the silicon chip in 1957. When political scientists William Ogburn and Dorothy Thomas studied this phenomenon, they found 148 cases of big ideas coming to many people at the same time and concluded that their list would grow longer with more research.

Having ideas is not the same thing as being creative. Creation is execution, not inspiration. Many people have ideas; few take the steps to make the thing they imagine. One of the best examples is the airplane. The brothers Orville and Wilbur Wright were not the first people to have the idea of building a flying machine, nor were they the first people to begin building one, but they were the first people to fly.

The Wright brothers' story begins in Germany's Rhinow Hills on Sunday, August 9, 1896. The sky stretched clean as a sheet, the moon chewed the sun in a partial solar eclipse, and a white shape soared between the peaks. It had the spoked wings of a bat and a crescent tail. A bearded man hung beneath: Otto Lilienthal, piloting a new glider, maneuvering by shifting his weight, aiming to create a powered flying machine. A gust of wind caught the glider and tilted it up. He swung his body but was unable to right it. His great white bat fell fifty feet, and Lilienthal thrashed in its jaws. His back was broken, and he died the next day. His last words were "Sacrifices must be made."

Orville and Wilbur Wright read the news at their Wright Cycle Company store in Dayton, Ohio. Lilienthal's sacrifice seemed senseless to them. No one should drive a vehicle he cannot steer, especially not in the sky.

Cycling was a new fashion in the 1890s. Bicycles are miracles of equilibrium. They are not easy to build or ride. When we cycle, we make constant adjustments to stay balanced. When we turn, we abandon this balance by steering and leaning, then recover it once our turn is complete. The problem of the bicycle is not motion; it is balance. Lilienthal's death showed the Wrights that the same was true of aircraft. In their book *The Early History of the Airplane,* the brothers wrote:

> The balancing of a flyer may seem, at first thought, to be a very
> simple matter, yet almost every experimenter had found in this
> one point which he could not satisfactorily master. Some experi-
> menters placed the center of gravity far below the wings. Like
> the pendulum, it tended to seek the lowest point; but also, like
> the pendulum, it tended to oscillate in a manner destructive of
> all stability. A more satisfactory system was that of arranging the

wings in the shape of a broad V, but in practice it had two serious defects: first, it tended to keep the machine oscillating; and second, its usefulness was restricted to calm air. Notwithstanding the known limitations of this principle, it had been embodied in almost every prominent flying machine that had been built. We reached the conclusion that a flyer founded upon it might be of interest from a scientific point of view, but could be of no value in a practical way.

In the same book, Wilbur added: "When this one feature has been worked out the age of flying machines will have arrived, for all other difficulties are of minor importance."

This observation set the Wright brothers on the path to the world's first flight. They saw an airplane as "a bicycle with wings." The problem of the aircraft is not flying: like the bicycle, it is balance. Otto Lilienthal died because he succeeded at the first and failed at the second.

The Wrights solved the problem by studying birds. A bird is buffeted by wind when it glides. It balances by raising one wingtip and lowering the other. The wind turns the wings like sails on a windmill until the bird regains equilibrium. Wilbur again:

To mention all the things the bird must constantly keep in mind in order to fly securely through the air would take a very considerable treatise. If I take a piece of paper, and after placing it parallel with the ground, quickly let it fall, it will not settle steadily down as a staid, sensible piece of paper ought to do, but it insists on contravening every recognized rule of decorum, turning over and darting hither and thither in the most erratic manner, much after the style of an untrained horse. Yet this is the style of steed that men must learn to manage before flying can become an everyday sport. The bird has learned this art of equilibrium, and learned it so thoroughly that its skill is not apparent to our sight. We only learn to appreciate it when we try to imitate it.

That is, when we try to fly a horse.

These were the Wrights' first mental steps. *Problem:* Balance a bucking aircraft. *Solution:* Imitate gliding birds.

The next problem was how to reproduce a bird's balance mechanically. Their first solution required metal rods and gears. This caused the next problem: it was too heavy to fly. Wilbur discovered the solution in the Wrights' bicycle shop while playing with a long, thin cardboard box that had once contained an inner tube—something roughly the same size and shape as a box of tin foil or Saran Wrap. When Wilbur twisted the box, one corner dipped slightly and the other rose by the same amount. It was a motion similar to a gliding bird's wingtips, but it used so little force that it could be achieved with cables. The distinctive double wings on the brothers' airplanes were based on this box; they called the twisting that made the tips go up and down "wing warping."

As young boys, the Wrights had loved to make and fly kites—"a sport to which we had devoted so much attention that we were regarded as experts." Despite their fascination, they stopped during their teenage years because it was "unbecoming to boys of our ages." And yet, twenty years later, Wilbur found himself cycling through Dayton as fast as he could with a five-foot kite across his handlebars. He had built it with wings that warped to prove the idea worked. He was hurrying to show it to Orville. The brothers had completed their second step.

And so it continued. The Wright brothers' great inventive leap was not a great mental leap. Despite its extraordinary outcome, their story is a litany of little steps.

For example, they spent two years trying to make Wilbur's kite big enough to carry a pilot before discovering that the aerodynamic data they were using was worthless.

"Having set out with absolute faith in the existing scientific data," they wrote, "we were driven to doubt one thing after another, till finally, after two years of experiment, we cast it all aside, and decided to rely entirely upon our own investigations."

The Wrights had started flying as a hobby and with little interest in "the scientific side of it." But they were ingenious and easily intrigued. By the time they realized that all the published data was wrong—"little better than guesswork"—they had also discovered what knowledge was needed to design wings that would fly. In 1901, they built a bicycle-mounted test platform to simulate airplanes in flight, then a belt-driven wind tunnel they used to create their own data. Many of the results surprised them—their findings, they wrote, were "so anomalous that we were almost ready to doubt our own measurements."

But they eventually concluded that *everybody else's* measurements were wrong. One of the biggest sources of error was the Smeaton coefficient, a number developed by eighteenth-century engineer John Smeaton to determine the relationship between wing size and lift. Smeaton's number was 0.005. The Wrights calculated that the correct figure was actually 0.0033. Wings needed to be much bigger than anybody had realized if an airplane was ever going to fly.

The Wrights used the same data to design propellers. Propellers had been built for boats but never for aircraft. Just as the brothers thought of an airplane as a bicycle that flew, they thought of a propeller as a wing that rotated. The lessons from their wind tunnel enabled them to design a near-perfect propeller on their first attempt. Modern propellers are only marginally better.

The Wrights' aircraft are the best evidence that they took steps, not leaps. Their glider of 1900 looked like their kite of 1899. Their glider of 1901 looked like their glider of 1900 but with a few new elements. Their glider of 1902 was their glider of 1901, bigger and with a rudder. Their 1903 *Flyer*—the aircraft that flew from Kitty Hawk's sands—was their 1902 glider made bigger again with propellers and an engine added. Orville and Wilbur Wright did not leap into the sky. They walked there one step at a time.

Thinking might make planes and phones, but surely art flows from soul to eye? Karl Duncker's mental steps may apply to the calculation of engineering, but do they also describe the majesty of art? To answer this question, we return to a Berlin on the brink of war.

On November 1, 1913, Franz Kluxen entered Berlin's Galerie Der Sturm to buy a painting. Kluxen was one of Germany's foremost collectors of modern art. He owned works by Marc Chagall, August Macke, Franz Marc, and a dozen Picassos. On this day another artist caught his eye—a controversial figure pushing painting to become ever more unreal: Wassily Kandinsky. The picture Kluxen bought was an abstract of contorting shapes and penetrating lines dominated by blues, browns, reds, and greens called *Bild mit weißem Rand,* or *Painting with White Border.*

A few months before Kluxen walked up to the finished painting in Berlin, Kandinsky had walked up to its blank canvas in Munich with a single piece of charcoal in his hand. The canvas was covered in a white paint made from five layers of zinc, chalk, and lead. Kandinsky had specified the paint precisely. He forbade artificial chalk made from gypsum and demanded more expensive natural chalk made from fossilized cells a hundred million years old.

Kandinsky drew a picture with the charcoal. Then he mixed paints using as many as ten pigments per color—his purple was made of white, vermilion, black, green, two yellows, and three blues, for example—and brushed them on in layers from lightest to darkest without pausing or missing a stroke. The picture covered thirty square feet, but Kandinsky finished it quickly. This speed and certainty created an impression of spontaneity. It was as if he awoke that morning and rushed to record a vanishing fragment of dream.

Art is the mastery of making appearance deceive. Kandinsky spent five months planning every stroke of his apparently spontaneous painting and years developing the method and theory that took him to it. Kandinsky was a Russian immigrant living in Germany. He visited his native Moscow in the fall of 1912 just as the First Balkan War began. To Russia's south, the Balkan League of Serbia, Greece, Bulgaria, and Montenegro was attacking Turkey, then called the Ottoman Empire. It was a brief, brutal war that started at the time of Kandinsky's trip and finished as he completed *Painting with White Border,* in May 1913. He returned to Germany packing a problem: how to paint the emotion of the moment—the "extremely powerful impressions I had experienced in Moscow—or more correctly, of Moscow itself."

He started by painting a sketch in oils he called *Mascau,* later renamed *Sketch 1 for Painting with White Border.* It was a constricted thicket of velveteen green with cadmium red accents and dark hemming lines. A trio of black curves oozed toward the top left corner, evoking the three-horse sled called a troika, a common Kandinsky motif and a symbol used by other Russians, including Nikolai Gogol, to represent their nation's divinity.

His second sketch, barely different, diffused the lines until they were more stain than stroke—in his words, "dissolving the colors and forms." More sketches followed. Kandinsky burnished his picture on paper, card, and canvas. He scrawled in pencil, mapping which colors would go where using letters and words. He brushed some studies with watercolor, others with gouache—a blend of gum and pigment halfway between watercolor and oil—and India ink. He crayoned. He made twenty sketches, each no more than one or two steps different from the last. The process took five months. The twenty-first picture— Kandinsky's finished work—is very similar to the first. *Painting with White Border* is the old friend you run into after a few years. *Sketch 1* is how the friend used to look. But vast differences hide beneath the surface of each piece. They tell the true story of artistic creation.

The green ground of *Sketch 1* is a mix of seven colors: green, umber, ocher, black, yellow, blue, and white. At the painting's center, Kandinsky first applied a yellow made from five colors: cadmium yellow, yellow ocher, red ocher, yellow lake, and chalk. Then, when the yellow was dry, he painted it over with green. These steps were not artistic: the canvas of *Sketch 1* had already been used, and Kandinsky had to cover an existing painting. He did such a good job that it was not until almost a hundred years later, after the advent of infrared imaging, that a team of conservators working for New York's Guggenheim Museum, which owns *Painting with White Border,* and Washington, D.C.'s Phillips Collection, which owns *Sketch 1,* discovered that there was a picture beneath the picture.

Once he had prepared the canvas, Kandinsky continued *Sketch 1* by layering colors from dark to light, rearranging and repainting the picture many times as he worked. This is partly visible from a close inspection of his brushstrokes and has been fully exposed by X-ray, which undresses a painting layer by layer. An X-ray of *Sketch 1* shows a blur: Kandinsky reworked the image so many times that only a few elements of the finished piece can be seen. He painted over almost everything on the canvas in fits of iteration that lasted until he solved his first problem: how to capture "the extremely powerful impressions I had experienced in Moscow."

When *Sketch 1* was complete, Kandinsky identified remaining problems one at a time. He rotated the image from portrait to landscape, softened the colors, and changed the ground from dark green to luminous white. One sketch shows twenty variations of the troika as Kandinsky tuned its curves like strings on a cello. And then there was the eponymous white border:

I made slow progress with the white edge. My sketches did little to help, that is, the individual forms became clear within me—and yet, I could still not bring myself to paint the picture. It tormented me. After several weeks, I would bring out the sketches again, and still I felt unprepared. It is only over the years that I have learned to exercise patience in such moments and not smash the picture over my knee.

Thus, it was not until after nearly five months that I was sitting in the twilight looking at the second large-scale study, when it suddenly dawned on me what was missing—the white edge. Since this white edge proved the solution to the picture, I named the whole picture after it.

With this final problem solved, Kandinsky ordered the canvas. When he first touched it with his charcoal, he knew exactly what he was about to make. Where an X-ray of *Sketch 1* shows a blur of painted work and rework, an X-ray of *Painting with White Border* is exactly like the painting itself. This is how we know he did not hesitate. After five months and twenty steps, Kandinsky was ready to paint.

The twenty steps are only part of the story. Kandinsky's journey did not begin with *Sketch 1,* and it did not end with *Painting with White Border.* His first works, painted in 1904, were colorful, realistic landscapes. His last, painted in 1944, were atonal, geometric abstracts. His first and last pictures look wholly unalike, but everything Kandinsky painted in the intervening years was a small step along the road that unites them. *Painting with White Border* marks a slight move toward more abstract images and is part of Kandinsky's transition from dark to light. Even in a lifetime of art, creation is a continuum.

As Karl Duncker showed, all creation, whether painting, plane, or phone, has the same foundation: gradual steps where a problem leads to a solution that leads to a problem. Creating is the result of thinking like walking. Left foot, problem. Right foot, solution. Repeat until you arrive. It is not the size of your strides that determines your success but how many you take.

CHAPTER 3

EXPECT ADVERSITY

1 | JUDAH

One summer night in 1994, a five-year-old named Jennifer crept downstairs to tell her mother her ear hurt. Jennifer's pediatrician prescribed eardrops. The pain got worse. One side of her face bulged. The pediatrician doubled her dose. The swelling grew. X-rays revealed nothing. The lump got bigger than a baseball. Jennifer glowed with fever, her head inflated, she lost weight. Surgeons removed the lump. It came back. They took half of Jennifer's jaw. Still the lump returned. It was removed again. It came a fourth time, reaching toward her skull to kill her. Medicine did not work. Jennifer's one chance was radiation. Nobody knew if it would affect the tumor. Everybody knew it would stop half her face from growing. Children with that condition often kill themselves.

As Jennifer's parents contemplated their choice, her doctor heard rumors of a researcher with a controversial theory that tumors create their own blood supply. This man said growths like Jennifer's could be destroyed by cutting off their access to blood. Very few people believed him, and his approach was so experimental that it was practically quackery. The man's name was Judah Folkman.

Jennifer's doctor told her parents about this unproven theory. He warned them that Folkman was a controversial man with a mixed reputation, possibly more fantasist than scientist. Jennifer's parents felt they had little to lose. "Fantasy" is just another word for hope with long odds. It is better than no hope. Jennifer's father signed a consent form and put his daughter's life in Judah Folkman's hands.

Folkman prescribed injections of a new, unproven drug. Jennifer's father, a machinist, gave her the shots. Her mother, who worked at a grocery store, held her. For weeks they stuck needles into Jennifer's arm, over her tear-soaked cries of protest. Folkman's shots made her worse. They boiled the disfigured, dying little girl in fever and terrified her with visions. Neighbors heard her screaming during the night and remembered her in their prayers.

Folkman called his theory *angiogenesis*—Latin for "growth of new blood vessels." He had conceived it more than thirty years earlier when one of his experiments went wrong. He was an enlisted man, a surgeon required to spend time in the navy researching new ways to store blood on long voyages. To see what methods might work, he built a maze of tubes that circulated blood substitutes through a rabbit gland and injected the gland with the fastest-growing things he knew of: cancer cells from a mouse. He expected the cells to either grow or die. But something else happened. The cells grew as big as dots on dice, then stopped. They were still alive; when Folkman put them back in the mice, they swelled into deadly tumors. Here was mystery. Why would cancer stop on a gland but kill in a mouse?

Folkman noticed that the tumors in the mice were full of blood and the tumors on the glands were not. In the mice, new blood vessels reached out, greeting the tumors, feeding and growing them.

Other navy lab scientists found this mildly interesting. Judah Folkman thought it was life-changing. He felt sure he had discovered something important. What if the tumors were creating these new vessels, weaving themselves a bloody web in which to grow? What if you could stop that from happening? Would it kill the tumors?

Folkman was a surgeon. Wrist-deep in living flesh, surgeons see things lab scientists do not. To a surgeon, a tumor is a wet red mess, like fat on a steak. To a scientist, it is dry and white, like a cauliflower. "I had seen and handled cancers, and they were hot and red and bloody," Folkman said. "And so when critics would say, 'Well, we don't see any blood vessels in these tumors,' I knew they were looking at tumors that had been taken out. All the blood was drained. They were specimens."

After leaving the navy, Folkman joined City Hospital in Boston. His lab was tiny, and the only natural light that dribbled in was from windows near the high ceiling.

He worked alone for years. When he finally recruited a team, it consisted of one medical student and one undergraduate. They worked nights and weekends on a debut paper about how blood vessels depend on cell fragments called platelets. It was published in *Nature* in 1969.

After that, Folkman's work was rejected. *Cell Biology, Experimental Cell Research,* and the *British Journal of Cancer* refused to print his papers on the connections between tumors and blood. His requests for grants were denied. Reviewers said his conclusions went beyond the data, that what he saw in his lab would not be seen in patients, and that his experiments were poorly designed. Some called him crazy.

In the 1960s and '70s, no one in cancer cared about blood. All the glory went to tumor killers wielding radiation and poison. Doctors targeted malignant cells as if they were marauding armies and attacked them with treatments inspired by war. Chemotherapy was developed from the chemical weapons of World War I; radiation resembled the nuclear weapons of World War II. Folkman imagined cancer as a disease of regeneration, not degeneration—a condition caused by the body growing, unlike most other illnesses, which are caused by the body decaying or failing. He did not picture tumors as invaders. He thought they were naturally communicating cells, having what his first research assistant, Michael Gimbrone, called a "dynamic dialogue" with the body. Folkman was convinced that he could stop this communication and make tumors die of natural causes.

One reason Folkman faced skepticism was that he was a surgeon. Scientists had little respect for surgeons. A surgeon's place was in the butcher's shop of the operating room, not the library of the laboratory. But Folkman said that seeing cancer in living people helped his work. He once rushed to the lab inspired by a patient whose ovarian cancer had spread beyond her ovaries. During the surgical procedure to save her, he had found a large tumor full of blood orbited by small white tumors that had not yet signaled for a supply. He thought life was confirming his ideas even though all the experts were rejecting them.

And their rejection was fierce. At best, Folkman's talks were met with apathy. At worst, audiences walked out when it was his turn to speak, leaving him facing an empty room. A member of one grant committee wrote that he was "working on dirt." Another said he was on "a hopeless search." A professor at Yale called him "a charlatan." Researchers were advised not to join his lab. Members of the board of Boston Children's Hospital, where he had been surgeon in chief, worried that Folkman was damaging their hospital's reputation. They cut his salary in half and forced him to quit performing surgery. One day in 1981, he repaired the deformed throat of a newborn baby girl, scrubbed out, and was never allowed to operate again.

The attacks from outside Folkman's lab were matched by disappointments within it. Trying to prove his hypothesis meant monotonous experiments, most of which were unsuccessful. He put a sign on his wall excusing his lack of progress. It said: "Innovation is a series of repetitive failures."

One Saturday in November 1985, Folkman's researcher Donald Ingber found fungus contaminating one of his experiments. This is not uncommon in laboratories. Scientists follow a strict protocol: they throw contaminated experiments away. Ingber did not do this. He examined the blood vessels growing in the petri dish. The fungus was forcing them to retreat.

Ingber and Folkman experimented with the fungus, watching as it blocked the growth of blood vessels in culture dishes, then chicken embryos, then mice.

Enter Jennifer. Folkman tried to exorcise her tumor to prove his crackpot theory of angiogenesis. As Jennifer twisted and screamed under the torture of Folkman's "treatment," her family came to realize why Folkman could not get published, funded, or perform surgery. It was the same reason other scientists called him a crazy charlatan on a hopeless search, walked out of his talks, said he was working on dirt, and told researchers to avoid him.

The reason was that his idea was new.

After the first few weeks of agonizing treatment, Jennifer's fevers cooled. Her hallucinations passed. The lump in her head shrank away until it left her forever. Her jaw grew back. She was a pretty little girl again. Judah Folkman had saved her life.

2 | FAIL

There are no shortcuts to creation. The path is one of many steps, neither straight nor winding but in the shape of a maze.

Judah Folkman walked the maze. It is easy to enter and difficult to stay.

Creation is not a moment of inspiration but a lifetime of endurance. The drawers of the world are full of things begun. Unfinished sketches, pieces of invention, incomplete product ideas, notebooks with half-formulated hypotheses, abandoned patents, partial manuscripts. Creating is more monotony than adventure. It is early mornings and late nights: long hours doing work that will likely fail or be deleted or erased—a process without progress that must be repeated daily for years. Beginning is hard, but continuing is harder. Those who seek a glamorous life should not pursue art, science, innovation, invention, or anything else that needs new. Creation is a long journey where most turns are wrong and most ends are dead. The most important thing creators do is work. The most important thing they don't do is quit.

The only way to be productive is to produce when the product is bad. Bad is the path to good. Until he saved Jennifer's life, Folkman described his work as "a series of repetitive failures." Those failures did not come easily. There were gory experiments with rabbit eyes, chicken embryos, and puppy intestines. Some ideas needed vast quantities of cow cartilage, others gallons of mouse urine. Many experiments had to be repeated many times. Some went wrong and were thrown away. Some went right but yielded unhelpful results. Much of the work took nights and weekends. Long periods of effort produced nothing. Folkman once wondered about the difference between futility and tenacity and came to a conclusion that became his mantra: "If your idea succeeds, everybody says you're persistent. If it doesn't succeed, you're stubborn."

Folkman saved more lives after Jennifer's. Angiogenesis became an important theory in the treatment of cancer. Doctors and scientists regarded Folkman as more than persistent—he was lauded as a genius. But he received that distinction only after he proved his hypothesis. Surely either he had been a genius all along or he was no genius at all?

Donald Ingber's fungus was not the miraculous coincidence it might seem to be. Endurance often finds fortune. Folkman and his team worked for years to discover ways to culture blood vessels, to test for blocking agents, and to understand the nature of tumor growth. Ingber was a brilliant scientist, working on a Saturday, prepared for chance. In any other lab, the fungus would have been thrown out. In other labs, doing different research, it almost certainly already had been. That event was a culmination, not a revelation. Luck favors work.

We enter creation's maze with problems at every turn. Folkman's beginning, working alone in a badly lit lab barely bigger than a toll-booth, was not auspicious. Neither were his first experiments. He started with more questions than answers. We will, too. Some we ask ourselves. Some are asked by others. We will not know the answers or even how to find them. Creation demands belief beyond reason. Our foothold is faith—in ourselves, in our dream, in our odds of success, and in the cumulative, compound, creative power of work. Folkman had no reason to know he was right and countless reasons to believe he was wrong—many of them provided by his peers. He continued because of faith.

Faith is how we face failure. Not faith in a higher power—although we may choose that, too—but faith that there is a way forward. Creators redefine failure. Failure is not final. It carries no judgment and yields no conclusions. The word comes from the Latin *fallere,* to deceive. Failure is deceit. It aims to defeat us. We must not be fooled. Failure is lesson, not loss; it is gain, not shame. A journey of a thousand miles ends with a single step. Is every other step a failure?

Stephen Wolfram, scientist, author, and entrepreneur, is best known for his geeky software program Mathematica. In addition to writing books and code, he obsessively gathers information about his life. He has amassed what he says is "one of the world's largest collections of personal data." He knows how many e-mails he has sent since 1989, how many meetings he has had since 2000, how many phone calls he has made since 2003, and how many steps he has taken since 2010. He knows these things precisely. Since 2002 he has logged every key he has ever pressed on his computer's keyboard. He made over one hundred million keystrokes in the ten years between 2002 and 2012 and was surprised to find that the key he pressed most often was Delete. He had used it more than seven million times: he erased seven out of every hundred characters he typed, a year and a half of writing, then deleting.

Wolfram's measurement includes around two hundred thousand e-mails. He found he deleted most often when he was writing for publication. This is true for professional writers, too. Stephen King, for example, has published more than eighty books, most of them fiction. He says he writes two thousand words a day. Between the beginning of 1980 and the end of 1999, he published thirty-nine new books, totaling more than five million words. But writing two thousand words a day for twenty years yields *fourteen* million words: King must erase almost two words for every one he keeps. He says, "That DELETE key is on your machine for a good reason."

Where do Stephen King's deleted words go? They are not all lost to rephrasing. One of King's most popular books is a novel called *The Stand,* published in 1978. The finished manuscript, submitted after he had made all his deletions, was, he says, "twelve hundred pages long and weighed twelve pounds, the same weight as the sort of bowling ball I favor."

His publishers were worried that such a long book would not sell, so King made more deletions: three hundred pages' worth. But his most telling revelation is that he might never have traveled that far: around the halfway point in the writing, after more than five hundred single-spaced pages, King got stuck: "If I'd had two or even three hundred pages I would have abandoned *The Stand* and gone on to something else—God knows I had done it before. But five hundred pages was too great an investment, both in time and in creative energy."

King will throw away three hundred single-spaced typewritten pages, about sixty thousand words, which will have taken him more than a month to write, if he feels they are not good enough.

Success is the culmination of many failures. When James Dyson, an inventor, finds a problem, he immediately builds something that does not solve it, an approach he calls "make, break, make, break." What the world calls a failure the engineer calls a prototype. From Dyson's website:

There's a misconception that invention is about having a great idea, tinkering with it in the tool shed for a few days, then appearing with the finished design. In fact, it's usually a far longer and iterative process—trying something over and over, changing one small variable at a time. Trial and error.

Dyson describes himself as "just an ordinary person. I get angry about things that don't work." The thing that made him so angry it changed his life was a vacuum cleaner that lost suction as its bag filled. He was thinking about it as he drove past a factory with a dust extractor that works based on a principal called "cyclonic separation." Cyclonic separators, or cyclones, spin air in a spiral and move anything else—like dust and dirt—around until it eventually drops down. This is is how Dorothy got to Oz:

> The north and south winds met where the house stood, and made it the exact center of the cyclone. In the middle of a cyclone the air is generally still, but the great pressure of the wind on every side of the house raised it up higher and higher, until it was at the very top of the cyclone. The little girl gave a cry of amazement and looked about her. The cyclone had set the house down very gently—for a cyclone—in the midst of a country of marvelous beauty.

The beauty of dust extraction by cyclone is simple: there is no filter to clog, which means nothing reduces the suction. Filters were the reason most vacuum cleaners sucked—or, rather, did not. Dyson's idea was equally simple: make a vacuum cleaner that used a cyclone instead of sucking dust and air through a filter.

Cyclone math is *not* simple—it combines fluid mechanics to describe the movement of air with particle transport equations to predict the behavior of dust. Dyson did not waste much time on this math. Like the Wright brothers, he made an observation, then went straight to making. And the first thing he made—out of cardboard and a disassembled vacuum cleaner—did not work. Neither did the second, third, or fourth.

Dyson faced many problems. He had to make the world's smallest cyclone. It had to be capable of extracting house dust particles about a millionth of a meter wide. And he had to make it suitable for home use and mass production.

It took more than five *thousand* prototypes, constructed over five years, to create a working cyclone-based vacuum cleaner. He says, "I'm a huge failure because I made 5,126 mistakes." And, on another occasion:

> I wanted to give up almost every day. A lot of people give up when
> the world seems to be against them, but that's the point when you
> should push a little harder. I use the analogy of running a race. It
> seems as though you can't carry on, but if you just get through the
> pain barrier, you'll see the end and be okay. Often, just around the
> corner is where the solution will happen.

Dyson's solution was—eventually—a working cyclone-based vacuum cleaner that created a multibillion-dollar business and a personal fortune of more than $5 billion.

Judah Folkman's observation that "innovation is a series of repetitive failures" applies to every field of creation and every creator. Nothing good is created the first time. The step-by-step approach to problem solving Karl Duncker observed does not apply only to *forward* movements like Kandinsky's sketches. Some steps go *backward*. But persistence turns everything into progress. Writer Linda Rubright's definition of "Iterative Process" is "Total fail. Repeat." Creators must be willing to fail and repeat until they find the step that arrives. Samuel Beckett said it best: "Try again. Fail again. Fail better."

Failure is not wasteful but useful. Time spent failing is time spent well. Wandering creation's maze is never a waste of time. Only leaving it is.

A Hungarian psychology professor once wrote to famous creators, asking them to be interviewed for a book he was writing. One of the most interesting things about his project was how many people said no.

Management writer Peter Drucker: "One of the secrets of productivity (in which I believe whereas I do not believe in creativity) is to have a VERY BIG waste paper basket to take care of ALL invitations such as yours—productivity in my experience consists of NOT doing anything that helps the work of other people but to spend all one's time on the work the Good Lord has fitted one to do, and to do well."

Secretary to novelist Saul Bellow: "Mr. Bellow informed me that he remains creative in the second half of life, at least in part, because he does not allow himself to be a part of other people's 'studies.'"

Photographer Richard Avedon: "Sorry—too little time left."

Secretary to composer György Ligeti: "He is creative and, because of this, totally overworked. Therefore, the very reason you wish to study his creative process is also the reason why he (unfortunately) does not have time to help you in this study. He would also like to add that he cannot answer your letter personally because he is trying desperately to finish a Violin Concerto which will be premiered in the Fall."

The professor contacted 275 creative people. A third of them said no. Their reason was lack of time. A third said nothing. We can assume their reason for not even saying no was also lack of time and possibly lack of a secretary.

Time is the raw material of creation. Wipe away the magic and myth of creating and all that remains is work: the work of becoming expert through study and practice, the work of finding solutions to problems and then problems with those solutions, the work of trial and error, the work of thinking and perfecting, the work of *creating*. Creating consumes. It is all day, every day. It knows neither weekends nor vacations. It is not when we feel like it. It is habit, compulsion, obsession, and vocation. The common thread that links creators is how they spend their time. No matter what you read, no matter what they claim, nearly all creators spend nearly all their time on the work of creation. There are few overnight successes and many up-all-night successes.

Saying no has more creative power than ideas, insights, and talent combined. Saying no guards time, the thread from which we weave our creations. The math of time is simple: you have less than you think and need more than you know.

We are not taught to say no. We are taught *not* to say no. No is rude. No is a rebuff, a rebuttal, a minor act of verbal violence. No is for drugs and strangers with candy.

But consider the Hungarian professor: famous, distinguished, politely and personally requesting a small amount of time from people who had already found creative success. And two-thirds of them declined, in most cases saying nothing or having someone else say no for them, wasting not even a minute to reply.

Creators do not ask how much time something takes but how much creation it costs. This interview, this letter, this trip to the movies, this dinner with friends, this party, this last day of summer. How much less will I create unless I say no? A sketch? A stanza? A paragraph? An experiment? Twenty lines of code? The answer is always the same: yes makes less. We do not have enough time as it is. There are groceries to buy, gas tanks to fill, families to love, and day jobs to do.

People who create know this. They know the world is all strangers with candy. They know how to say no, and they know how to suffer the consequences. Charles Dickens, rejecting an invitation from a friend:

"It is only half an hour"—"It is only an afternoon"—"It is only
an evening," people say to me over and over again; but they don't

know that it is impossible to command one's self sometimes to any stipulated and set disposal of five minutes—or that the mere consciousness of an engagement will sometime worry a whole day. Who ever is devoted to an art must be content to deliver himself wholly up to it, and to find his recompense in it. I am grieved if you suspect me of not wanting to see you, but I can't help it; I must go in my way whether or no.

No makes us aloof, boring, impolite, unfriendly, selfish, antisocial, uncaring, lonely, and an arsenal of other insults. But no is the button that keeps us on.

Failure is often followed by rejection.

In 1846, large numbers of women and babies were dying during childbirth in Vienna. The cause of death was puerperal fever, a disease that swells then kills its victims. Vienna's General Hospital had two maternity clinics. Mothers and newborns were dying in only one of them. Pregnant women waited outside the hospital, begging not to be taken to the deadly clinic, often giving birth in the street if they were refused. More women and babies survived labor in the street than in the clinic. All the deaths came at the hands of doctors. In the other clinic, midwives delivered the babies.

Vienna General was a teaching hospital where doctors learned their trade by cutting up cadavers. They often delivered babies after dissecting corpses. One of the doctors, a Hungarian named Ignaz Semmelweis, started to wonder if the puerperal fever was somehow being carried from the corpses to the women in labor. Most of his peers thought the question preposterous. Carl Edvard Marius Levy, a Danish obstetrician, for instance, wrote that Semmelweis's "beliefs are too unclear, his observations too volatile, his experiences too uncertain, for the deduction of scientific results." Levy was offended by the lack of theory behind Semmelweis's work. Semmelweis speculated that some kind of organic matter was being transferred from the morgue to the mothers, but he did not know what it was. Levy said this made the whole idea unsatisfactory from a "scientific point of view."

But, from a *clinical* point of view, Semmelweis had convincing data to support his hypothesis. At a time when doctors did not scrub in or out of the operating room, and were so proud of the blood on their gowns that they let it build up throughout their careers, Semmelweis persuaded the doctors of Vienna to *wash their hands* before delivering babies, and the results were immediate. In April 1847, 57 women died giving birth in Vienna General's deadly First Clinic—18 percent of all patients. In the middle of May, Semmelweis introduced handwashing. In June, 6 women died, a death rate of 2 percent, the same as the untroubled Second Clinic. The death rate stayed low, and in some months fell to zero. In the following two years, Semmelweis saved the lives of around 500 women, and an unknown number of children.

This was not enough to overcome the skepticism. Charles Delucena Meigs, an American obstetrician, typified the outrage. He told his students that a doctor's hands could not possibly carry disease because doctors are gentlemen and "a gentleman's hands are clean."

Semmelweis did not know why hand-washing before delivery saved lives—he only knew that it did. And if you do not know *why* something saves lives, why do it? For Levy, Meigs, and Semmelweis's other "gentlemen" contemporaries, preventing the deaths of thousands of women and their babies was not reason enough.

As the medical community rejected Semmelweis's ideas, his morale and behavior declined. He had been a rising star at the hospital until he proposed hand-washing. After a few years, he lost his job and started showing signs of mental illness. He was lured to a lunatic asylum, put in a straitjacket, and beaten. He died two weeks later. Few attended his funeral. Without Semmelweis's supervision, the doctors at Vienna General Hospital stopped washing their hands. The death rate for women and babies at the maternity clinic rose by 600 percent.

Even in a field as apparently empirical and scientific as medicine, even when the results are as fundamental as life not death, and even when the creation is as simple as asking people to wash their hands, creators may not be welcome.

Why? Because powerful antibodies of the status quo mass against change. When you bring something truly new to the world, brace. Having an impact is not usually a pleasant experience. Sometimes the hardest part of creating is not having an idea but saving an idea, ideally while also saving yourself.

Semmelweis's idea challenged two millennia of medical dogma. Since the time of Hippocrates, doctors had been trained in humorism: the belief that the body is made up of four fluids, or humors: black bile, yellow bile, phlegm, and blood. Humorism lives in our language today. In Latin, black bile is *melan chole*. People with too much of it were said to suffer from *melancholy*. Too much yellow bile, *chole*, made a person irritable, or *choleric*. An excess of blood, *sanguis*, made them optimistic, or *sanguine*. Phlegm made them stoic, or *phlegmatic*. Good health meant these humors were in balance. Disease and disability came from imbalances caused by inhaling vapors or "bad air," an idea known as "miasma theory." Diseases were treated by removing blood. In the nineteenth century, doctors removed blood by placing leeches on their patients' bodies, a treatment called "hirudotherapy." The leeches attached themselves to the patient's skin using a sucker, behind which lay a three-bladed, propeller-shaped jaw. Once the sucker was in place, the leech latched on by biting, injected anesthetic and blood thinners into the patient; then it sucked the patient's blood. Once full, it dropped off to begin digestion. The process took up to two hours. It was important to wait. If the leech was removed prematurely, it would vomit into the patient's open wound.

Semmelweis's idea that puerperal fever might be carried by doctors from corpses to patients and could therefore be prevented by hand-washing contradicted the ancient trinity of humorism, miasma, and hirudotherapy. How could hygiene impact health when disease was generated spontaneously inside the body?

As Semmelweis lay dying, another creator, Louis Pasteur, answered this question. Where Semmelweis pointed to the number of women who did not die and expected common sense to prevail, Pasteur used carefully designed experiments to advance what became known as "germ theory." He produced incontrovertible evidence to show that living microorganisms caused many diseases. Pasteur was well aware of the controversial nature of his theory and possibly also of the hostile rejection that proponents like Semmelweis had suffered. Humorism's true believers had been fighting rumors about germs for centuries. Pasteur was meticulous with his evidence, persistent with his claims, and eventually convinced most of Europe. Semmelweis's clinical results hinted at the truth, but they were not enough to overcome two thousand years of belief in something else. A new idea needs much better evidence than an old one, as some of our best thinkers have pointed out.

David Hume: "A wise man proportions his belief to the evidence."

Pierre-Simon Laplace: "The weight of evidence for an extraordinary claim must be proportioned to its strangeness."

Marcello Truzzi: "An extraordinary claim requires extraordinary proof."

Carl Sagan: "Extraordinary claims require extraordinary evidence."

Prevailing ideas are fortified by incumbency and familiarity, no matter how ridiculous they may seem later. They can only be changed by people ready to meet rejection with evidence, patience, and stamina. Semmelweis believed that saving hundreds of women was enough.

One reason for Semmelweis's collapse was that he did not expect such a good idea to be so soundly rejected and was shocked at the vicious and sometimes personal attacks. But creation is the infiltration of the old by the new, a stone in the shoe of the status quo, and this makes creators threats, at least to some. As a consequence, creation is seldom welcome.

Still, Semmelweis's surprise is typical. The most common misconception about creation is that good ideas are celebrated—partly because of something that happened in Concord, Massachusetts, in 1855.

5 | BETTER MOUSETRAPS

In a long, flowing hand that joined words as well as letters, Ralph Waldo Emerson wrote in his journal, "If a man has good corn, or wood, or boards, or pigs, to sell, or can make better chairs or knives, crucibles, or church organs, than anybody else, you will find a broad, hard beaten road to his house, though it be in the woods." By 1889, several years after Emerson's death, the line was being misquoted as "If a man can write a better book, preach a better sermon, or make a better mousetrap than his neighbor, though he builds his house in the woods, the world will make a beaten path to his door." Later it was changed again, to "Build a better mousetrap and the world will beat a path to your door," and became famous.

These words did more than cause a misunderstanding about the popularity of new things in general. Many people take them literally, and as a result, the mousetrap has become one of the most frequently patented and reinvented devices in America. Around four hundred applications for mousetrap patents are made every year. About forty patents are granted. More than five thousand mousetrap patents have been issued in total—so many that the U.S. Patent and Trademark Office has thirty-nine subclasses for mousetraps, including "Impaling," "Choking or Squeezing," and "Electrocuting and Explosive." Independent inventors hold nearly all mousetrap patents. Almost all of them cite the quotation they believe is Emerson's. But the world does not beat a path to their door. Fewer than twenty of the five thousand mousetrap patents have ever made any money.

The saying was not intended to inspire better mousetraps. Rather, a better mousetrap inspired it. Emerson could not have written it: he died before commercial mousetraps were invented. I know the story well in part because my great-grandfather, who lived at the same time as Emerson, made his living as a rat catcher. His principal tools were dogs: Jack Russell terriers, a relatively new breed in those days and one developed specifically for hunting vermin. Other mouse-trapping techniques included cats—actually less effective than dogs, despite their reputation—as well as cages and drowning. This changed in the late 1880s, when an inventor from Illinois named William C. Hooker created the first mass-production mousetrap. And not long after that, my family's trade changed, too. There was little demand for rat catchers when people could buy cheap traps. Hooker's trap is the one we know today: a spring-loaded bar released by a trigger when a mouse takes the bait. This is the "better mousetrap" referred to in the 1889 revision of Emerson's words. It does not need building: William Hooker has already built it.

Hooker's "snap trap" was perfected within a few years. It was cheap, easy, and effective. It remains the dominant design today. It traps a quarter of a billion mice a year, outsells all its competitors combined by a factor of two to one, and costs less than a dollar. Almost all of the five thousand mousetraps created since Hooker's have been rejected.

The idea that creators are hailed as heroes is as wrong today as it was when Emerson did not write it. Emerson's actual point was about what he called "common fame"—the success a person has in their community if they provide valuable goods or services. If he were writing today, Emerson might have said, "Open the best coffee shop in town and your neighbors will wait in line for a cup." He is not exhorting us to invent an alternative to coffee.

The mistaken belief that the world awaits a better mousetrap has yielded more than mousetraps. It has given rise to an industry of predators. Businesses called "invention promotion companies" advertise on television and radio and in newspapers and magazines, promising to evaluate people's ideas, patent them, and sell them to manufacturers and retailers. They charge an initial fee of hundreds of dollars for "evaluation." The evaluation almost always concludes that a person's idea is patentable and valuable. Then the companies demand thousands of dollars for legal and marketing services. Inventors are made to feel that their idea has been specially selected. They are given the impression that the invention promotion company will be investing its own time and money in their idea so it can earn royalties. In fact, the companies make all their money from the up-front fees. They have little success marketing inventions or helping inventors.

In 1999, the U.S. federal government intervened to protect the "Nation's most precious natural resource: the independent inventor." The American Inventor's Protection Act was signed into law by President Clinton, and the Federal Trade Commission filed lawsuits against invention promotion companies operating under names including National Idea Center, American Invention Associates, National Idea Network, National Invention Network, and Eureka Solutions International. In a moment of knowing poetry, the FTC called its program Project Mousetrap.

One company, Davison & Associates, settled with the FTC by making a payment of $11 million and promising not to misrepresent its services. The company has since changed its name to Davison Design. It has an amusement park–like "factory" called Inventionland, complete with a castle, pirate ship, and tree house hidden behind a false bookcase in its offices in O'Hara, Pennsylvania. Inventionland is staffed by employees called "Inventionmen." Their creations include a pan for making meatballs, a rail for storing flip-flops, and clothes for dogs. Many of their inventions are based on Davison's own ideas, even though they are billed as "client products."

Inventionland is where we find the truth about better mousetraps. The FTC settlement forced Davison to disclose how many of its clients make money. According to the company's November 2012 report, an average of eleven thousand people a year sign its agreements. Of these, three make a profit. In the twenty-three years between Davison's founding, in 1989, and 2012, twenty-seven people have made money using the company's services, barely more than one a year. How much money? Davison has to disclose its prices. Its customers multiplied by its prices equals sales of $45 million a year. Davison says the money it makes on sales of its customers' products is 0.001 percent of its revenue and that this represents a 10 percent royalty on what its customers make. If this is correct, Davison makes $450 a year from royalties and Davison's customers all added together receive a total of $4,050 a year for the $45 million a year they spend on the company's services—less than one dollar returned for every ten thousand invested.

Davison offers to sign up more than sixty thousand ideas a year. This alone should make an inventor suspicious—and it might but for the myth of the better mousetrap. Unfortunately, anyone who loves your idea the first time they hear it either loves you or wants something. What to expect when you're inventing is rejection. Build a better mousetrap and the world will not beat a path to your door. You must beat a path to the world.

6 | THE MOST DECISIVE OF DENIALS

Rejection hurts, but it is not the worst thing that can happen. On February 22, 1911, Gaston Hervieu grasped the railing on the first platform of the Eiffel Tower and looked down. He was almost two hundred feet above Paris. The people watching him from the ground looked smaller than his fingernails.

Hervieu was an inventor of parachutes and airships. In 1906, he was part of a team that tried to reach the North Pole by airship; in 1909, he developed a parachute to slow the descent of aircraft. Hervieu had climbed the Eiffel Tower to test a new emergency parachute for pilots. He checked the wind, took a nervous breath, and began the test. His parachute opened as soon as it cleared the platform. The silk filled with air, making a hemisphere in the sky, then sailed safely to the ground. A photographer from the Dutch newsweekly *Het Leven* captured the moment: a figure descending gracefully, silhouetted beneath the tower's northwestern arch, with the watching crowd and the Palais du Trocadéro in the background.

There was a catch, though. Hervieu did not make the jump himself; he used a 160-pound test dummy instead. To most people this seemed prudent, but for at least one man, it was an outrage. Franz Reichelt was an Austrian tailor who was developing a parachute of his own. He denounced Hervieu's use of a dummy as a "sham" and, one year later, on the morning of Sunday, February 4, 1912, arrived at the Eiffel Tower to conduct an experiment of his own.

Reichelt had made sure his test would be publicized. Photographers, journalists, and a cameraman from the Pathé news service were all waiting to meet him. He posed for pictures, doffed his black beret, and then made an announcement that took most people by surprise. He would not be using a dummy or even a safety harness. He said, "I am so convinced my device will work properly that I will jump myself."

Gaston Hervieu, who had come to the Eiffel Tower to watch Reichelt's test, tried to stop him. Hervieu claimed there were technical reasons why Reichelt's parachute would not work. The two men had a heated discussion until, finally, Reichelt turned away and walked to the tower's staircase. As he began his ascent, he looked back and said, "My parachute will give your arguments the most decisive of denials."

Hervieu had carried a parachute and a dummy up the 360 steps to the tower's first floor, but Reichelt carried nothing: he was *wearing* his parachute, just as a pilot would if he were about to leap from a crashing airplane. Reichelt's description of the concept appeared in news stories the following day: "My invention has nothing in common with similar devices. It is partly waterproof fabric and partly pure silk. The first serves as clothing and adapts to the body like ordinary clothes; the second consists of a parachute which is folded behind the pilot like a backpack."

Two assistants were waiting for him when he reached the top of the staircase. They set a chair on a table so that Reichelt could stand above the railing and jump. For more than a minute, he stayed with one foot on the chair and the other on the railing, looking down, checking the wind, and making last-minute adjustments. It was below freezing in Paris, and his breaths came out as steamy plumes. Then he stepped off the railing, into the void.

A photographer from *Het Leven* waited beneath the tower's northwestern arch, ready to take a picture exactly like the one of Hervieu's test, only showing a living man, not a dummy.

That picture does show a living man, but it is different from the one taken a year earlier in another way. Where the photograph of Hervieu's test shows a perfect parachute, the photograph of Reichelt's test shows a blur like a broken umbrella. The broken umbrella is Reichelt. His "parachute" did not work. It was a suit of clothes intended to turn the person wearing it into something like a flying squirrel. Large silk sheets connected Reichelt's arms to his ankles, and a hood stood above his head. Reichelt fell for four seconds, accelerating constantly, until he hit the ground at sixty miles an hour, making a cloud of frost and dust and a dent six inches deep. He was killed on impact.

Modern parachutes use 700 square feet of fabric and should be deployed only above 250 feet; Reichelt's parachute used less than 350 square feet of fabric, and he deployed it at 187 feet. He had neither the surface area nor the altitude needed to make a successful jump; this was why Hervieu had tried to stop him.

Hervieu was not the only one who had told Reichelt that his parachute suit would not work; it had also been rejected by experts at the Aéro-Club de France, who had written, "The surface of your device is too small. You will break your neck."

Reichelt ignored all these rejections until the only thing left to reject him was reality. And, as physicist Richard Feynman said seventy-four years later, "For a successful technology, reality must take precedence over public relations, for nature cannot be fooled."

Dramatic ending aside, Reichelt's story is the story of most would-be creators. We hear about creation's few wins and never know its many losses. Tales like Ignaz Semmelweis's are carefully chosen cherries. Much of their power comes from dramatic irony: we know the creator will be vindicated in the end. This can make creators seem like heroes and rejecters seem like villains. But rejecters are nearly always sincere. They want to stop wrong and dangerous thinking. They believe they are right, and they usually are. If Reichelt had landed, we would read his story differently. Reichelt would seem like a hero, Hervieu a jealous rival, and the Aéro-Club de France a group of out-of-touch obstructionists. But only the outcome would be different. The motives of Reichelt's rejecters would be unchanged.

Rejection has value.

7 | THE REJECTION REFLEX

Judah Folkman was rejected for decades. His grants were denied, his papers returned, his audiences hostile. He endured lawsuits, demotion, innuendo, and insult. But he was a charming man. He inspired his researchers, was always available to patients, and told his wife he loved her every day. Folkman was not rejected because he was bad or because his ideas were bad. He was rejected because rejection is a natural consequence of new.

Why? Because we fear new as much as we need it.

In the 1950s, two psychologists, Jacob Getzels and Phillip Jackson, studied a group of high school students. All the students had higher-than-average IQs, but Getzels and Jackson found that the most creative students tended to have lower IQs than the least creative students. As part of the study, the students wrote brief autobiographies. One higher-IQ student wrote:

My autobiography is neither interesting or exciting and I see very little reason for writing it. However I shall attempt to write a certain amount of material which would be constructive. I was born May 8, 1943, in Atlanta, Georgia, USA. I am descended from a long line of ancestry which is mostly Scotch and English, with a few exceptions here and there. Most of my recent ancestors have lived in the southern US for a good while, though some are from New York. After being born, I remained in Georgia six weeks, after which I moved to Fairfax, Virginia. During my four-year stay there, I had few adventures of any kind.

One highly creative student wrote:

In 1943 I was born. I have been living without interruption ever since. My parents are my mother and father—an arrangement I have found increasingly convenient over the years. My father is a physician and surgeon—at least that's what the sign on his office door says. Of course, he's not anymore for Dad's past the age where men ought to enjoy the rest of his life. He retired from

Mercy Hospital Christmas before last. Got a fountain-pen for 27 years of service.

The difference between these two passages is typical of the differences the study found between the high-IQ children and the highly creative children. The highly creative children were funnier, more playful, less predictable, and less conventional than the high-IQ children. This was no surprise. The surprise was the teachers: they liked the high-IQ children, but they did *not* like the creative children. Getzels and Jackson were amazed. They had expected the opposite, because their experiment had revealed something else: the highly creative children were delivering academic results as good as or better than the high-IQ children—a performance far better than the twenty-three-point deficit in their IQ scores would predict. If you believed in IQ scores—and all the teachers in this school did—the highly creative children were beating the odds. But, even though the highly creative children were star performers who were exceeding expectations, the teachers did not like them. They preferred the less creative children who were performing as expected.

This was not a freak result. It has been repeated many times, and it remains the same today. The vast majority—98 percent—of teachers say creating is so important that it should be taught daily, but when tested, they nearly always favor less creative children over more creative children.

The Getzels-Jackson effect is not restricted to schools, and it persists into adulthood. Decision makers and authority figures in business, science, and government all say they value creation, but when tested, they do not value creators.

Why? Because people who are more creative also tend to be more playful, unconventional, and unpredictable, and all of this makes them harder to control. No matter how much we say we value creation, deep down, most of us value control more. And so we fear change and favor familiarity. Rejecting is a reflex.

We do not only reject other people's creative instincts; often, we reject our own, too.

In one experiment, Dutch psychologist Eric Rietzschel asked people to score ideas based on how "feasible," "original," and "creative" they were, and then asked them which ones were "the best." The ideas people selected as "the best" ideas were almost always the ones they had scored as the least "creative."

When Rietzschel asked people to assess their own work, he got the same result: almost everybody thought their least "creative" ideas were their "best" ideas.

The findings are highly repeatable. Decades of data all show the same thing: even though we say we want creation, we tend to reject it.

8 | THE NATURE OF NO

The tendency to welcome new ideas in principle then reject them in practice is a feature, not a bug. Every species has its niche, and every niche has its risk and reward. The human race's niche is the niche of new. Our reward is rapid adaptation: we can change our tools faster than evolution can change our bodies. Our risk is that the footsteps of new lead into darkness. Creating something new may kill us; creating nothing new certainly will. This makes us creatures of contradiction: we need and fear change. No one is only progressive or only conservative. Each of us is both. And so we say we want new, then choose same.

Our innate drive for new would make us extinct if it were unrestrained. Everyone would die trying everything. The instinct to reject is evolution's solution to our problem of needing to make new while needing to take care.

We are wired to reject new things, or at least be suspicious of them. When we are in familiar situations, cells in our brain's seahorse-shaped core, the hippocampus, fire hundreds of times faster than they do when we are in new situations. The hippocampus is connected to two tiny balls of neurons called amygdalae—from the Greek for "almonds"— that drive our emotions. The connection between the hippocampus and the amygdalae is one reason that same feels good and new may not.

As our brains react, so do we. We swerve from what feels bad to what feels better. When something is new, our hippocampus finds few matching memories. It signals unfamiliarity to our amygdalae, which give us feelings of uncertainty. Uncertainty is an aversive state: we avoid it if we can. Psychologists can show this in experiments. Feelings of uncertainty bias us against new things, make us prefer familiarity, and stop us from recognizing creative ideas. This happens even when we value creation or think we are good at creating.

To make matters worse, we fear rejection, too. As anyone who has lost a lover knows, rejection hurts. We use phrases like "broken-hearted," "bruised egos," and "hurt feelings," because we feel physical pain when spurned—a word that, not coincidentally, comes from the Old English *spurnen*, "to kick." In 1958, psychologist Harry Harlow proved something Aristotle had proposed twenty-five hundred years earlier: we need love like we need air. In experiments no ethics committee would allow today, Harlow separated newborn monkeys from their mothers. The babies preferred a soft, cloth surrogate mother doll to a wire surrogate, even though the wire version delivered food. Monkeys deprived of a soft surrogate often died, despite having enough food and water. Harlow called his paper "The Nature of Love" and concluded that physical contact is more important than calories. His finding extends to humans. We would rather die hungry than lonely.

Our primal need for connection doubles the dilemma of novelty. We are biased against new experiences, but it is hard to admit that we feel this way, even to ourselves, because we also face social pressure to make positive statements about creative ideas. We know we should not suggest that being creative is bad. We may even self-identify as "creative." The bias against new is a prejudice a bit like sexism and racism: we know it is socially unacceptable to "dislike" creation, we sincerely believe we "like" creation, but when presented with a specific creative idea, we are more likely to reject it than we realize. And, when we present a creative idea to others, *they* are much more likely to reject it than *they* realize. It is human nature to say no to new.

Sexism and racism are famous prejudices. The bias against new is not. No one talks about *newism*. "Luddism," our closest word, is a misunderstanding. The Luddites—about whom much more later—were English weavers who destroyed automatic looms to protect their jobs during the late eighteenth and early nineteenth centuries. Although Luddism was, in the words of Thomas Pynchon, an effort to "deny the machine," the attack on new technology was incidental. The Luddites were not fighting against new. They were fighting for their livelihoods. Yet their name fills the void in our vocabulary where there is a fear without a noun.

The bias against new is no less real because it goes unnamed; if anything, its anonymity makes it worse. Labels make things visible. Women and racial minorities are not surprised by prejudice against them. The words "sexism" and "racism" signal that sexism and racism exist. Newism comes with no such warning. When companies, academies, and societies revere creation in public and then reject it in private, creators *are* surprised and wonder what they did wrong.

Boston Children's Hospital's rejection of Judah Folkman is typical. Children's is one of America's highest-ranked hospitals and part of Harvard University, the oldest institute of higher learning in the United States. The hospital houses more than a thousand scientists and has produced Nobel laureates and Lasker Award winners. It is a place where new ideas should be welcomed. Yet Children's punished Folkman for having a theory about cancer that his contemporaries found controversial. The hospital is proud of him now. But in 1981, when it stopped him from being a surgeon and reduced his pay, it was not. I chose Folkman's story because it shows how flowers are sometimes mistaken for weeds. The point is not that Boston Children's Hospital did something wrong. The point is that it did something normal.

What is *not* normal about Folkman's story is his tenacity. It is hard to withstand repeated rejection. But we cannot create unless we know what to do with no.

9 | ESCAPING THE MAZE

How do we escape this maze of rejection, failure, and distraction?

Rejection is a reflex that evolved to protect us. No matter what we may gain, our first reactions to new are suspicion, skepticism, and fear. This is the right response: most ideas are bad. Stephen Jay Gould: "A man does not attain the status of Galileo merely because he is persecuted; he must also be right."

Creators must expect rejection. The only way to avoid rejection is to avoid making anything new. Rejection is not a ticket to quit. It does not mean the work is bad. It does not mean *we* are bad. Rejection is about as personal as gravity.

At its best, rejection is information. It shows us what to do next. When Judah Folkman's early critics argued that he was seeing inflammation, not blood vessels, he designed experiments to exclude inflammation. Rejection is not persecution. Drain it of its poison and what remains may be useful.

Franz Reichelt, the parachutist who leapt to his death, did not listen to the lessons of rejection or failure. He not only ignored experts who pointed out the flaws in his design, he ignored his own data. He tested his parachute using dummies, and they crashed. He tested his parachute by jumping thirty feet into a haystack, and *he* crashed. He tested his parachute by jumping twenty feet *without* a haystack, and he crashed *and* broke his leg. Instead of changing his invention again and again until it worked, he clung to his bad idea in the face of all evidence and stopped thinking at the first solution he'd found.

The creation is not the creator. Great creators do not extend their belief in themselves to their work. A creation can be changed. The problem-solution loop never ends. Reichelt's loop ended soon after it began. His tragedy is a metaphor for the problem with leaps. He saw a problem and tried to solve it not with a series of steps but with a leap both literal and figurative. He was not an artist of new but a martyr to same.

Ignaz Semmelweis, the hand-washing obstetrician so vexed by rejection that he lost his job, then his life, missed a huge opportunity. Semmelweis had found something of world-changing importance: a link between cadavers and disease. His critics complained that he did not know what it was. He believed that saving lives was convincing enough. But it was not. If he had taken his rejection less personally and fought back by devoting himself to understanding more, he, not Louis Pasteur, might have discovered germs, and his contribution might have saved lives everywhere forever, instead of in one place for a few years.

Failure is a kind of rejection best done in private. The greatest creators are their own greatest critics. They look at their work even more deeply than other people will, and they test it against more exacting standards. They reject most of what they make either in part (as Stephen King does when he throws away two-thirds of his words) or in whole (as James Dyson does when he condemns yet another prototype) many times before anybody else gets to. The world is already inclined to reject you. Do not give it more reasons than necessary. Never have a failure in public that you could have in private. Private failures are faster, cheaper, and less painful.

Our nature does not help us. In addition to the discomfort with ambiguity that pushes us to want to find solutions quickly, there is also the problem of pride. Pride and its opposite, shame, can make us fearful of failure and resentful of rejection. Our ego does not want to hear no. We want to be right the first time, make a quick buck, be an overnight success. The creativity myth, with its roots in genius, aha! moments, and other magic, appeals to the part of us that wants to win without work, get without sweat, make no mistakes. None of these things are possible. Do not take pride in your work. Earn it.

We can learn a lot from what people do when they get lost in *real* mazes—along backcountry hiking routes, on terrain crossed with old trails, and in other places where losing your way can be deadly. This is only somewhat metaphorical. Whether we are creating or walking, we are trying to get somewhere by taking steps and making choices.

William Syrotuck analyzed 229 cases of people who became lost, 25 of whom died. He found that when we are lost, most of us act the same way. First, we deny that we are going in the wrong direction. Then, as the realization that we are in trouble seeps in, we press on, hoping chance will lead us. We are least likely to do the thing that is most likely to save us: turn around. We know our path is wrong, yet we rush along it, compelled to save face, to resolve the ambiguity, to achieve the goal. Pride propels us. Shame stops us from saving ourselves.

Great creators know that the best step forward is often a step back—to scrutinize, analyze, and assess, to find faults and flaws, to challenge and to change. You cannot escape a maze if you only move forward. Sometimes the path ahead is behind.

Rejection educates. Failure teaches. Both hurt. Only distraction comforts. And of these, only distraction can lead to destruction. Rejection and failure can nourish us, but wasted time is a tiny death. What determines whether we will succeed as creators is not how intelligent we are, how talented we are, or how hard we work, but how we respond to the adversity of creation.

Why is changing the world so hard? Because the world does not want to change.

CHAPTER 4

HOW WE SEE

1 | ROBIN

June 1979 was a cold, wet month in Western Australia. The worst day was Monday, June 11. An inch of rain fell, blown by a mean wind that turned windows into drums. Behind a loud window in Perth, a man with a silver beard and bolo tie looked through his microscope and saw something that would change the world.

Robin Warren was a pathologist at the Royal Perth Hospital. What he saw were bacteria in a patient's stomach. Scientists had known that bacteria could not grow in the stomach since the beginning of bacteriology. Stomachs are acidic, so they had to be sterile. The bacteria Warren saw were curved like croissants. They flattened the brushy surface of the stomach's lining. Warren could see them at magnifications of one hundred, but his colleagues could not. He showed them images magnified one thousand times, then some taken with a high-power electron microscope. They eventually saw the bacteria but not the point. Only Warren thought the discovery meant something, although he did not know what.

He did not rush to judgment, like Ignaz Semmelweis, nor did he disregard possible objections, like Franz Reichelt, nor did he allow his rejection instinct to delete the gleam of something new. Warren was a quieter, shier man than Judah Folkman, but his response to being the only person in the lab who thought he had seen something significant was Folkmanesque. He believed what he saw, he believed it might be important, and he would not be dissuaded. In his report of that day's biopsy, he wrote: "It contains numerous bacteria. They appear to be actively growing and not a contaminant. I am not sure of the significance of these unusual findings, but further investigation may be worthwhile."

Having seen the bacteria once, he saw them often. They were in one out of every three stomachs. The dogma of the sterile stomach said bacteria could not live in the gut. No one else had ever seen bacteria there. "The apparent absence of any previous report was given to me as one of the main reasons why they could not be there at all," he said.

Warren collected samples of the bacteria that were not there for two years, until he found someone who believed him.

Barry Marshall was a newly hired gastroenterologist who needed a research project. Warren, like all pathologists, did not see patients. He worked with samples clinicians gave him, most of which were from ulcers and lesions. This made seeing the bacteria harder: activity around the wounds added noise. Marshall agreed to send Warren biopsies from ulcer-free sites, and the two men started to collaborate.

Within a year, Warren and Marshall had one hundred clean samples. They found that 90 percent of patients who had the bacteria had ulcers. *Every* patient with a duodenal ulcer—erosion in the lining of the acidic passage at the start of the intestine, immediately after the stomach—had the bacteria.

The two men tried and failed to grow the bacteria in the hospital's lab. For six months they started with live samples and ended with nothing.

Then, during Easter 1982, a drug-resistant superbug contaminated the hospital and overwhelmed its lab. Warren and Marshall's samples were forgotten for five days. The lab staff had been discarding the samples after two. The bacteria grew in the extra three. All they had needed was more time.

The bacterium was new. It was eventually given the name *Helicobacter pylori,* or *H. pylori* for short. Warren and Marshall wrote about their discovery in a 1984 letter to the *Lancet,* one of the world's highest-impact medical journals. They concluded that the bacteria, which "appeared to be a new species, were present in almost all patients with active chronic gastritis, duodenal ulcer, or gastric ulcer and thus may be an important factor in these diseases."

The *Lancet*'s editor, Ian Munro, could not find any reviewers who agreed. Everybody knew bacteria could not grow in the stomach. The results had to be wrong. Fortunately for Warren and Marshall—and all of us—Ian Munro was no ordinary journal editor; he was a radical thinker who, among other things, campaigned for human rights, nuclear disarmament, and medicine for the poor. In an unusual and impactful moment of science as it should be, Munro published the letter over the objections of his reviewers, even adding a note saying, "If the authors' hypothesis should prove valid this work is very important indeed."

Warren and Marshall went on to show that *H. pylori* causes ulcers. Others, building on their work, learned how to cure ulcers by killing *H. pylori* with antibiotics. In 2004, Warren and Marshall won the Nobel Prize "for their discovery of the bacterium *Helicobacter pylori* and its role in gastritis and peptic ulcer disease." We now know that there are hundreds of species of bacteria in the stomach and that, among other things, they play an essential role in keeping the digestive system stable.

There is something strange about this story.

What Robin Warren saw on that cold, wet Monday was not something no one had seen. It was something everyone had seen. The only thing he did that no one else had done was believe it. By 1979, Warren had spent seventeen years mastering the complex science of pathology—the careful preservation and examination of human tissue—especially analyzing stomach biopsies. These became common in the 1970s, after the invention of the flexible endoscope—a tube with a light, camera, and cutting tool that doctors could feed down the throat of a patient and use to extract tissue. Before this, most samples were either from whole stomachs that had been removed or from cadavers. These were difficult to process. Information was lost while the samples were made ready for analysis. It was these bad samples that had led to every doctor and scientist being taught that bacteria do not live in the stomach. Warren said, "This was taken as so obvious as to barely rate a mention." His biopsies told a different story.

"As my knowledge of medicine and then pathology increased, I found that there are often exceptions to 'known facts,'" he said.

Also, "I preferred to believe my eyes, not the medical textbooks or the medical fraternity."

He makes it sound simple. Yet flexible endoscopes were being used all over the world. Thousands of pathologists were looking at stomach biopsies. *H. pylori* was staring them all in the face. But they saw dogma, not bacteria.

In June 1979, the month Warren first noticed *H. pylori,* a group of American scientists published a paper about an epidemic of stomach disease among participants in a research study. The volunteers were healthy at the start of the project; then half of them became ill with stomach pain, followed by a loss of stomach acidity. The illness was almost certainly infectious. The scientists tested the patients' blood and stomach fluid. They looked for a virus—because they knew bacteria could not grow in the stomach—and they did not find one. Their conclusion was: "We have been unable to isolate an infectious agent, nor have we been able to establish a viral or bacterial cause." These were not beginners; they were led by a decorated professor of medicine who was also editor in chief of the journal *Gastroenterology*. After Warren and Marshall's work was published, these scientists revisited their biopsies. *H. pylori* was clearly visible. They had seen it and not seen it. Their patients had been suffering from an acute infection of the bacteria. One of the scientists said, "Failing to discover *H. pylori* was my biggest mistake."

In 1967, Susumo Ito, a professor at Harvard Medical School, had biopsied his own stomach and used an electron microscope to take a perfect photograph of *H. pylori*. It appeared labeled as a "spirillum," but without further comment or attempt at identification, in that year's American Physiological Society *Handbook of Physiology*. Tens of thousands of scientists saw the picture. None of them saw *H. pylori*.

In 1940, Harvard researcher Stone Freedberg had found *H. pylori* in more than a third of ulcer patients. His supervisor told him he was wrong and made him stop his research.

Only Robin Warren believed and would not be dissuaded. He maintained a lonely vigil over *H. pylori* for two years, until Marshall arrived.

H. pylori has now been found in medical literature dating back to 1875. When Robin Warren discovered it, it had been seen and not believed for 104 years.

2 | WHAT YOU SEE IS NOT WHAT YOU GET

H. pylori's tiny boomerangs hid in plain sight for more than a century because of a problem called "inattentional blindness." The name comes from perception psychologists Arien Mack and Irvin Rock, but the best definition comes from novelist Douglas Adams:

> Something that we can't see, or don't see, or our brain doesn't let us see, because we think that it's somebody else's problem. The

brain just edits it out; it's like a blind spot. If you look at it directly you won't see it unless you know precisely what it is. It relies on people's natural predisposition not to see anything they don't want to, weren't expecting or can't explain.

Adams uses this definition in his book *Life, the Universe and Everything* in a scene where nobody notices that an alien spacecraft has landed in the middle of a cricket match. The story is comic, but the concept is real: the brain is the secret censor of the senses. It takes steps between sensing and thinking that we do not notice.

The path from eye to mind is long. Each eye has two optic nerves, one for the right half of the brain and one for the left. They travel as far back as they can possibly go—to an outside layer at the back of the brain called the visual cortex. Touch the back of your head, and your hand is next to the part of your brain that connects to your eyes. The visual cortex compresses what your eyes see by a factor of ten, then passes the information to the center of the brain, the striatum. The information is compressed again, this time by a factor of three hundred, as it travels to its next stop, the basal nuclei at the striatum's core. This is where we discover what the eyes have seen and decide what to do about it. Only one three-thousandth of what is rendered on the retina gets this far. The brain selects what gets through by adding prior knowledge and making assumptions about how things behave. It subtracts what does not matter and what has not changed. It determines what we will and will not know. This preprocessing is powerful. What the brain adds seems real. What it subtracts may as well not exist.

This is why it is a bad idea to have a phone conversation while driving. Using a phone halves the amount of sensory information that enters our mind. Our eyes stare at the same things for the same length of time, but our brain edits out most of the information as unimportant. The information may be important for driving, but our brain preprocesses it based on what is important for our phone call. This does not happen when we are listening to the radio, because the radio does not expect us to talk back. It does not happen when we are talking to a passenger, because the passenger is in the same space we are. But studies show that when we talk on the phone, we get inattentional blindness. The conversation is our problem. That child unexpectedly crossing the street in front of us is somebody else's problem. Our brain does not let us see her. As Douglas Adams described it, our brain blinds our mind to the unusual.

This is also true when we are walking. In one study, researchers put a clown on a unicycle in the path of pedestrians. The researchers asked people who walked past the clown if they had noticed anything unusual. Everybody saw him unless they had been on their cell phone. Three out of every four people who had been using their phone did not see the clown. They looked back in astonishment, unable to believe they had missed him. They had looked straight at him but had not registered his presence. The unicycling clown crossed their paths but not their minds.

Harvard researchers Trafton Drew and Jeremy Wolfe did a similar experiment with radiologists by adding a picture of a gorilla to X-rays of lungs. An X-ray section of a lung looks like a black-and-white picture of a bowl of miso soup. As radiologists flick through images, they see progressive slices of the lungs, as if they are looking deeper into the soup. In Drew and Wolfe's images, a crudely cut out black-and-white picture of a man in a gorilla suit had been added to the top right corner of some of the layers, as if he were floating on his back there. The radiologists saw the tiny nodules that indicated whether each lung was cancerous, but almost all of them missed the gorilla, even though it was shaking its fist at them and would have occupied as much space as a matchbox if it had actually been present in the lungs. Each radiologist who did not see the gorilla looked at it for about half a second.

Inattentional blindness is not an experimental effect. In 2004, a forty-three-year-old woman went to the emergency room suffering from fainting and other symptoms. The doctors suspected heart and lung problems, so they put a catheter into her body using a guidewire that went from her thigh to her chest. The doctors forgot to remove the guidewire. Five days passed before anybody found it. The woman spent all that time in intensive care, where she had three X-rays and a CT scan as attempts were made to stabilize her. A dozen doctors looked at the images. The guidewire in her chest—which, fortunately, did not contribute to her condition—was obvious on all of them, but nobody noticed it.

3 | OBVIOUS FACTS

When Robin Warren accepted his Nobel Prize, he quoted Sherlock Holmes: "There is nothing more deceptive than an obvious fact."

It was an "obvious fact" that bacteria do not live in the stomach, just as it was an "obvious fact" that emergency room doctors remember to remove guidewires and an "obvious fact" that there are no gorillas in pictures of lungs.

Radiologists are experts in seeing. Years of training and practice make what is invisible to us obvious to them. They can diagnose a disease after looking at a chest X-ray for a fifth of a second, the time it takes to make a single voluntary eye movement. If you or I were to look at an X-ray of a lung, we would scan the whole thing, searching for irregularities. This is also what novice radiologists do. But as they become more trained, they move their eyes less, until all they have to do is glance at a few locations for a few moments to find the information they need.

This is called "selective attention." It is a hallmark of expertise. "Expert" has the same Latin root as "experience." Aldous Huxley, writing in his 1932 book *Texts and Pretexts,* says: "Experience is a matter of sensibility and intuition, of seeing and hearing the significant things, of paying attention at the right moments, of understanding and coordinating."

Adriaan de Groot, a chess master and psychologist, studied expertise by showing a chess position to players of different ranks, including grandmasters and world champions, and asking them to think aloud as they considered their move. De Groot had two expectations. First, that better players would make better moves. Second, that better players would make more calculations. What he found surprised him.

The first thing he noticed was the same problem-solution loop undergraduates used to solve the candle problem, Apple used to design the iPhone, and the Wright brothers used to invent the airplane.

A chess expert's first step is to evaluate the problem. One master started like this:

"Difficult: this is my first impression. The second is that by actual numbers I should be badly off, but it is a pleasant position."

The second step is to think of a move:

"I can do a whole lot of things. Get my Rook into it, at the Pawns."

Each move is evaluated after it is generated:

"No, a touch of fantasy. Not worth much. No good. Maybe not so crazy."

De Groot discovered several things. First, unfamiliar problems are solved with slow loops that are easily verbalized. Second, everybody revisits and reevaluates some solutions. This is not indecision: each evaluation goes deeper.

What surprised De Groot was how the problem-solution loop differed between players of different ranks. He'd expected grandmasters to make the best moves, and they did. But he had thought that this would be due to more analysis. What he found was the opposite. Grandmasters evaluated fewer moves and reevaluated them less often than other players did. One grandmaster evaluated one move twice, then evaluated another and played it. It was the best possible move. This was generally true: despite evaluating fewer moves fewer times, four of the five grandmasters in the study made the best possible move. The other grandmaster made the second best possible move. Grandmasters did not consider any moves that were not in the top five best moves. Lower-ranked players considered moves as poor as the twenty-second best. The less expert the player, the more options they considered, the more evaluations they made, and the worse their eventual move was.

Less thinking led to better solutions. More thinking led to worse solutions. Was this evidence of genius and epiphany? Were grandmasters making their moves by inspiration?

No. De Groot noticed something odd as he listened to grandmasters thinking aloud. Here is a typical comment from a grandmaster:

"First impression: an isolated Pawn; White has more freedom of movement."

Compare this to a master, a skilled player just one rank down, talking about the same position:

"The first thing that strikes me is the weakness of the Black King's wing, particularly the weakness at KB6. Only after that a general picture of the position. Finally, the complications in the center are rather striking: possibilities for exchange in connection with the loose Bishop on K7. Still later: my Pawn on QN2 is *en prise*."

En prise means the piece is vulnerable to being taken—this is the "isolated Pawn" the grandmaster mentioned first. We do not need to understand chess to see that the grandmaster came to an instant conclusion where the master took more time. De Groot hypothesized that the grandmasters' "remarks represent but a fraction of what has, in reality, been perceived. By far the largest part of what the subject 'sees' remains unsaid."

Experts do not think less. They think more efficiently. The practiced brain eliminates poor solutions so quickly that they barely reach the attention of the conscious mind.

De Groot showed this with another experiment. Grandmaster Max Euwe (a world champion), a master (De Groot himself, with his wife acting as experimenter), an expert-level player, and a class-level player were shown a position for five seconds, then asked to reconstruct it and think about a move. For Euwe, the grandmaster, this was trivial—he reconstructed the board easily. De Groot, the master, put nearly all the pieces in the right place but argued with his wife because he thought she had made a mistake setting up the board: "Is there really a Black Knight on KB2? That would be rather curious!" The expert-level player remembered three-quarters of the board; the class-level player, less than half.

Was grandmaster Euwe a genius? Did he have a photographic memory? No. As De Groot suspected, forcing Euwe to reconstruct the position showed he was thinking in fast loops:

"First impression: awfully rotten position, strong compressed attack by White. The order in which I saw the pieces was about King on K1, Knight on Q2, White Queen on QB3, Queen on K2, Pawns on K3 and his on K4, White Rook on Q8, White Knight on QN4, Rook on QN5—that funny Rook that doesn't do anything—Knight on KB2, Bishop on KB1, Rook on KR1, Pawns on KR2 and KN3. I didn't look at the other side very much, but I presume there is another Pawn on QR2. The rest for White: King on KN8, Rook on KB8, Pawns on KB7, KN7, KR7, and QR7, QN7."

In the five seconds he was given to look at the board, Euwe had seen the pieces in priority order, understood the logic of the position, and started reasoning about his move. He was doing ordinary thinking extraordinarily fast. His speed came from experience. It enabled him to see similarities to other games and connections between pieces. He did not *remember* the positions of the pieces—he *inferred* them. For example, he reconstructed the position that De Groot had assumed was a mistake by inference and without doubt: "Another piece is on

KB2—the King was completely closed in—that must be a Knight then."

The position reminded him of another game—"there's a vague recollection of a Fine-Flohr game in the back of my mind"—and all the similarities he saw gave him "a certain feeling of being familiar with this sort of situation." Experience enabled him to find a solution almost instantly.

Grandmasters have not been grandmasters forever. When they were masters, they played like masters, evaluating more moves more times. When they were expert-level players, they played like expert-level players—evaluating even more moves even more times. Because they have evaluated so many moves and accumulated so much experience, grandmasters can pay very selective attention to a game. The expert's first impression is not a first impression at all. It is the latest in a series of millions.

Creating is thinking. Attention is what we think *about*. The more we experience, the less we think—whether in chess, radiography, painting, science, or anything else. Expertise is efficiency: experts use fewer problem-solution loops because experts do not consider unlikely solutions.

Which means "selective attention" is another way of saying "obvious facts." As Robin Warren and Sherlock Holmes remind us, obvious facts can deceive. They are all we will see with the blindness of inattention. Developing expertise is essential, but it can block us from seeing the unexpected.

Becoming an expert is only the first step to becoming creative. As we are about to find out, the second step is surprising, confusing, and perhaps even intimidating: it is becoming a beginner.

In 1960, twelve elderly Japanese Americans waited at a gate in San Francisco International Airport. It had been nineteen years since the Japanese navy had attacked America's Pacific Fleet in Pearl Harbor. After the attack, these men and women were imprisoned in stables at a horse track in San Bruno. Three years later, the American government set each of them free with twenty-five dollars and a train ticket, then dropped atom bombs on Japan. They were Zen Buddhists and congregants of Sokoji, or the San Francisco Temple, which they had built in an abandoned synagogue near the Golden Gate Bridge in the calm before the war. They had continued to pay the mortgage while imprisoned. They were at the airport to meet their new priest.

As the sun rose, a silver-and-white Japan Air Lines Pacific Courier arrived from Honolulu, where it had stopped for fuel on its twenty-four-hour journey from Tokyo. Passengers started down stairs behind its port-side wing. Only one traveler, a tiny man in robes, sandals, and socks, looked energetic: Shunryu Suzuki, the priest.

Suzuki arrived in America when it was on the cusp of the 1960s. Children of the war were coming of age, animosity toward Japan was becoming curiosity, and young San Franciscans had started visiting Sokoji to ask about Zen Buddhism. Suzuki gave them all the same answer: "I sit at 5:45 in the morning. Please join me."

It was an invitation to the seated meditation called *zazen* in Japanese and *dhyāna* in Sanskrit. People in India and East Asia had sat in quiet contemplation for thousands of years, but the practice was little known in America. The few Americans who had tried it used chairs. Suzuki made his students sit on the floor with their legs crossed, backs upright, and eyes half-open. If he suspected they were sleeping, he struck them with a stick called a *kyōsaku*.

The class grew throughout the 1960s. In 1970, Suzuki's American students published his teachings in a book. The next year, a little more than a decade after his arrival, he died. The book, *Zen Mind, Beginner's Mind,* was as small, modest, and inspiring as he was. His was American Buddhism's first voice. His book is still in print.

Beginner's mind, *shoshin* in Japanese, was the essence of Suzuki's teaching. He described it simply: "In the beginner's mind there are many possibilities, but in the expert's there are few."

In Zen, simple words can have deep meanings. Beginner's mind is not the mind of the beginner but the mind of the master. It is an attention beyond the selection and blindness of expertise, one that notices everything without assumption. Beginner's mind is not mystical or spiritual but practical. It is Edmond Albius looking at a flower, Wilbur and Orville Wright looking at a bird, Wassily Kandinsky looking at a canvas, Steve Jobs looking at a phone, Judah Folkman looking at a tumor, Robin Warren looking at a bacterium. It is seeing what is there instead of seeing what we think.

Nyogen Senzaki, one of the first Zen monks in America, explained beginner's mind with a story, or *kōan:*

Nan-in, a Japanese master, received a university professor who came to inquire about Zen.

Nan-in served tea. He poured his visitor's cup full, and then kept on pouring.

The professor watched the overflow until he no longer could restrain himself. "It is overfull. No more will go in!"

"Like this cup," Nan-in said, "you are full of your own opinions and speculations. How can I show you Zen unless you first empty your cup?"

David Foster Wallace made the same point with a joke:

There are these two young fish swimming along and they happen to meet an older fish swimming the other way, who nods at them and says, "Morning, boys. How's the water?" And the two young fish swim on for a bit, and then eventually one of them looks over at the other and goes "What the hell is water?"

Creation is attention. It is seeing new problems, noticing the unnoticed, finding inattentional blind spots. If, in retrospect, a discovery or invention seems so obvious we feel as if it was staring us in the face all along, we are probably right. The answer to the question "Why didn't I think of that?" is "beginner's mind."

Or as Suzuki writes in *Zen Mind, Beginner's Mind:* "The real secret of the arts is to always be a beginner."

To see the unexpected, expect nothing.

5 | STRUCTURE

While Shunryu Suzuki was teaching Eastern philosophy at Sokoji, Thomas Kuhn was teaching Western philosophy on the other side of San Francisco Bay, in Berkeley. Kuhn was recovering from a great disappointment. He had spent sixteen years at Harvard University, earned three degrees in physics, and become a member of the university's elite Society of Fellows, but had been denied a position as a tenured professor. He had come to California to rebuild his career.

Kuhn's problem was that he had changed his mind. His degrees were in physics, but while working on his PhD, he had developed an interest in philosophy, a subject for which he had passion but no training. He also taught an undergraduate course on the history of science. He was not a scientist, philosopher, or historian but some odd combination of all three. Harvard was not sure what to do with him, and, he soon discovered, neither was the University of California, which hired him as a professor of philosophy, then changed his role to include history. It was clear he was no longer a scientist. The rest was fog.

This change in Kuhn's path started one summer afternoon when he first read Aristotle's *Physics*. The conventional view was that the book laid a foundation for all the physics that followed, but Kuhn could not see it. For example, Aristotle says:

> Everything that is in locomotion is moved either by itself or by something else. In the case of things that are moved by themselves it is evident that the moved and the movement are together: for they contain within themselves their first movement, so that there is nothing in between. The motion of things that are moved by something else must proceed in one of four ways: pulling, pushing, carrying, and twirling. All forms of locomotion are reducible to these.

This is not imprecise Newtonian physics or incomplete Newtonian physics; it is not Newtonian physics at all. The more Kuhn read old science, the more he realized that it was not connected to the science that followed it. Science is not a continuum, he concluded, but something else.

So, Kuhn wondered, what are we to make of these old theories? Were they not science and the people who practiced them not scientists? Did Newtonian physics also cease to be scientific when Einsteinian physics replaced it? How does science move from one set of theories to another, if not by gradually building on the work of the past?

By 1962, after fifteen years of research, Kuhn had his answer. He published it in a book called *The Structure of Scientific Revolutions*. He proposed that science proceeded in a series of revolutions where ways of thinking changed completely. He called these ways of thinking "paradigms." A paradigm is stable for a time, and scientists work on proving things that the paradigm predicts, but eventually exceptions appear. Scientists treat the exceptions as unanswered questions at first, but if enough of them are discovered and the questions are important enough, their paradigm is thrown into "crisis." The crisis continues until a new paradigm emerges. Then the cycle begins again. In Kuhn's view, a new paradigm is not an improved version of its predecessor. New paradigms vanquish old paradigms altogether. This is why it is impossible to understand scientists like Aristotle through a modern lens: they were working in a paradigm that has since been overthrown by scientific revolutions.

Despite its obscure topic, Kuhn's book has sold more than a million copies and is one of the most cited works in the world. Science writer James Gleick called it "the most influential work of philosophy in the latter half of the 20th century."

Paradigms are a form of selective attention. What changes during one of Kuhn's "scientific revolutions" is what scientists see. In Kuhn's words: "During revolutions scientists see new and different things when looking with familiar instruments in places they have looked before. What were ducks in the scientist's world before the revolution are rabbits afterwards."

Robin Warren's "discovery" of the bacterium *H. pylori,* which occurred after Kuhn's book was published, may be the clearest example of scientists seeing what they expect, not what is there. After Warren, scientists looked at images they had looked at before and were amazed to see things they had not previously seen. Their expertise—the system of beliefs, experiences, and assumptions Kuhn calls a paradigm—had blinded them.

6 | THE LINE BETWEEN EYE AND MIND

Seeing is not the same as looking. Knowing changes what we see as much as seeing changes what we know—not in a metaphorical or metaphysical way but literally. People on cell phones did not see the unicycling clown. Radiologists did not see the gorilla. Generations of scientists did not see *H. pylori*. This is not because the mind plays tricks but because the mind *is* a trick. Seeing and believing evolved because making sense of the world enabled species to survive and reproduce. Later, we became conscious and creative and wanted more from our senses, but as soon they were good enough for survival and reproduction, they were good enough for everything. We may want to believe that we inhabit a stable, objective universe and that our senses and minds render it fully and accurately—that what we perceive is "real"—and we may need to believe this so we can feel sane enough and safe enough to get on with our lives, but it is not true. If we want to understand the world enough to change it, we must understand that our senses do not give us the whole picture. Neil deGrasse Tyson, speaking at the Salk Institute in 2006, said:

> There is so much praise for the human eye, but anyone who has seen the full breadth of the electromagnetic spectrum will recognize how blind we are. We cannot see magnetic fields, ionizing radiation, or radon. We cannot smell or taste carbon monoxide, carbon dioxide, or methane, but if we breathe them in we are dead.

We know that these things exist because we have developed tools that sense them. But whether we use senses or sensors or both, our perception will always be limited by what we can detect and how we understand it. The first limitation is obvious—we know our eyes cannot see without light, for example—but the second, understanding, is not. There is a line between eye and mind. Not everything makes it across.

Creating means opening this border: reshaping our understanding so we notice things we have not noticed before. They do not have to be big or extraordinary. David Foster Wallace told his joke about fish to introduce something apparently mundane:

> After work you have to get in your car and drive to the supermarket. The supermarket is very crowded. And the store is hideously lit and infused with soul-killing Muzak. It's pretty much the last place you want to be. And who are all these people in my way? Look at how repulsive most of them are, and how stupid and cowlike and dead-eyed and nonhuman they seem, or at how annoying and rude it is that people are talking loudly on cell phones. Look at how deeply and personally unfair this is. Thinking this way is my natural default setting. It's the automatic way that I experience the boring, frustrating, crowded parts of life.
>
> But there are totally different ways to think about these kinds of situations. You can choose to look differently at this fat, dead-eyed, over-made-up lady who just screamed at her kid in the checkout line. Maybe she's not usually like this. Maybe she's been up three straight nights holding the hand of a husband who is dying of bone cancer. If you really learn how to pay attention, it will actually be within your power to experience a crowded, consumer-hell type situation as not only meaningful, but sacred. You get to consciously decide what has meaning and what doesn't.

When we change what has meaning, we change what we see. Wallace offers an alternative paradigm for the line at the grocery store. The lady's appearance stays the same, but he sees her differently. His second interpretation—that her husband has bone cancer—is speculative and probably wrong, but it is not more speculative or more likely to be wrong than his first. It is probably closer to the truth: few of us are generally mean, but all of us have difficult days that make us look mean to strangers. Wallace directs his selective attention to select something else. He can do this because he recognizes that his "natural default" way of seeing is not his *only* way of seeing. It is a choice. His ability to choose to see ordinary things differently—"as not only meaningful, but sacred"—made him one of the greatest writers of his generation.

Beginner's mind and expertise sound like opposites, but they are not. Western philosophy has conditioned us to see things in opposing pairs—black and white, left and right, good and evil, yin and yang (as opposed to the original Chinese idea of yin-yang), beginner and expert—a paradigm called "dualism." We do not have to see things this way. We can see them as connected, not opposed. Beginner's mind is *connected to,* not *opposite to,* expertise because the greatest experts understand that they are working within the constraints of a paradigm and they know how those constraints arose. In science, for example, some constraints are the result of available tools and techniques. Robin Warren had developed enough expertise as a pathologist to know that the dogma of the sterile stomach predated the invention of the flexible endoscope and might be a wrong assumption caused by a lack of technology. Judah Folkman knew that assumptions about tumors were based on specimens, not surgery. The Wright brothers knew that the Smeaton coefficient for calculating the relationship between wing size and lift was an assumption developed in the eighteenth century that might be wrong. The greatest test of your expertise is how explicitly you understand your assumptions.

There are no true beginners. We start building paradigms as soon as we are born. We inherit some, we are taught some, and we infer some. When we first create, we are already David Foster Wallace's fish, swimming in a sea of assumptions we have not yet noticed. The final step of expertise is the first step to beginner's mind: knowing what you assume, why, and when to suspend your assumptions.

There is a problem with seeing things no one else sees. How do we know we are right? What's the difference between necessary confidence and dangerous certainty—between discovery and delusion?

In the summer of 1894, Percival Lowell looked through the telescope in his new observatory for the first time. He had already announced he was starting "an investigation into the condition of life on other worlds," with "strong reason to believe that we are on the eve of a pretty definite discovery in the matter."

Lowell watched the ice on the south pole of Mars melt in the summer heat. Other astronomers had seen straight lines crossing the Martian desert. As the ice melted, the lines changed color, becoming lighter in the south and darker in the north. As far as Lowell was concerned, there was only one possible explanation: the lines were artificial canals—an "amazing blue network on Mars that hints that one planet besides our own is actually inhabited now."

Lowell inspired a century of science fiction starring marauding Martians. Many matched Lowell's descriptions. For example, in *Under the Moons of Mars,* Edgar Rice Burroughs wrote, "The people had found it necessary to follow the receding waters until necessity had forced upon them their ultimate salvation, the so-called Martian canals."

Scientists were less convinced. One of Lowell's opponents was Alfred Wallace, known for his work on evolution. Wallace did not challenge Lowell's maps. The Lowell Observatory was one of the best in the world, and Wallace had no reason to doubt what Lowell saw. Instead Wallace attacked Lowell's conclusions with a list of logical flaws, including:

The totally inadequate water-supply for such worldwide irrigation; the extreme irrationality of constructing so vast a canal-system the waste from which, by evaporation, would use up ten times the probable supply; how the Martians could have lived before this great system was planned and executed; why they did not first utilize and render fertile the belt of land adjacent to the limits of the polar snows; the fact that the only intelligent and practical way to convey a limited quantity of water such great distances would be by a system of water-tight and air-tight tubes laid *under the ground;* and only a dense population with ample means of subsistence could possibly have constructed such gigantic works—even if they were likely to be of any use.

The argument was resolved in Wallace's favor in 1965 when NASA's *Mariner 4* spacecraft took pictures of Mars that showed, in the words of the mission's imaging engineer, a surface "like that of our own Moon, deeply cratered, and unchanged over time. No water, no canals, no life."

But there was still one mystery. Whenever other astronomers said they could not see canals on Mars, Lowell pointed out that he had a better observatory. This was largely true. Few people had access to Lowell's private observatory while he was alive, but after he died, astronomers were finally able to look through his telescope. Still, no one could see any canals. What had Lowell been seeing?

The answer turned out to be his own eyes. Lowell was not an experienced astronomer. He had mistakenly made the aperture of his telescope so small that it worked like an ophthalmoscope, the handheld device doctors use to shine light into the eyes of patients. The veins on Lowell's retina were reflecting onto the lens of his telescope's eyepiece. His maps of Martian canals are mirror images of the tree of blood vessels we all have on the backs of our eyes—as are the "spokes" he saw on Venus, the "cracks" he saw in Mercury, the "lines" he saw on Jupiter's moons, and the "tores" he saw on Saturn.

Lowell was looking at a projection from inside his head. No telescope is more powerful than the prejudice of the person looking through it. We can see what we expect when it does not exist just as well as we can ignore the unexpected when it does.

Seeing what we expect shares the same root as inattentional blindness. When we prime our eyes with preconception, we do not have beginner's mind. Lowell could have avoided his errors. A. E. Douglass, his assistant, pointed out the risk of making the telescope's aperture so small not long after Lowell started using it: "Perhaps the most harmful imperfection of the eye is within the lens. Under proper conditions it displays irregular circles and radial lines resembling a spider-web. These become visible when the pencil of light entering the eye is extremely minute."

Douglass tested his hypothesis by hanging white globes a mile from the observatory. When he looked at them through the telescope, he saw the same lines Lowell was mapping onto planets. Lowell's response was to fire Douglass—soon to become a distinguished astronomer—for disloyalty.

We can change direction when we take steps but not when we make leaps. Lowell made a *leap* when he said he would find life on Mars. He committed himself to canals, not truth. Robin Warren took a *step* when he said there were bacteria in the stomach. He made that modest note in his lab report: "I am not sure of the significance of these unusual findings, but further investigation may be worthwhile." Then he took more steps. They led to the Nobel Prize.

Warren did not lack confidence—he was not, for example, deterred by colleagues who said the bacteria were of no significance—but he did lack something Lowell had in abundance: certainty.

Confidence is belief in yourself. Certainty is belief in your beliefs. Confidence is a bridge. Certainty is a barricade.

Certainty is even easier to create than illusion. Our brains are electrochemical. The feeling of certainty, like any other feeling, comes from the electrochemistry in our heads. Chemical and electrical stimulation can make us feel certain. Ketamine, phencyclidine, and methamphetamine create feelings of certainty, as does applying electricity to the entorhinal cortex, a part of the brain a few inches behind the nose.

False certainty is common in everyday life. In a study of memory, cognitive psychologists Ulric Neisser and Nicole Harsch tested false certainty by asking students how they first heard about the explosion of the space shuttle *Challenger.* One student's answer was: "I was in my religion class and some people walked in and started talking about it. I didn't know any details except that it had exploded and the schoolteacher's students had all been watching which I thought was so sad."

Another response was: "I was sitting in my freshman dorm room with my roommate and we were watching TV. It came on a news flash and we were both totally shocked."

Both answers are from the same student. Neisser and Harsch first asked her the question the day after the event, then tracked her down two years later and asked the question again. She felt "absolutely certain" about the second answer.

Of the forty people in the *Challenger* study, twelve were wrong about everything they recalled, and most were wrong about most things. Thirty-three were sure they had never been asked the question before. There was no relationship between the subjects' feelings of certainty and their accuracy. Being wrong, even being *shown* we are wrong, does not stop us from feeling certain.

Nor does irrefutable evidence—in fact, there is no such thing. Everybody continued to believe their second, incorrect memory even when shown the answers they had handwritten the day after the explosion. One response: "I still think of it as the other way around."

Once we become certain, we can remain certain, even when the evidence that we are mistaken should be overwhelming. This unshakable certainty was first studied in 1954, when psychic and spiritualist Dorothy Martin said aliens had warned her that the world would be destroyed on December 21. Psychologists Leon Festinger, Stanley Schachter, Henry Riecken, and others posed as believers, joined her group of followers, and watched what happened when her prophecy did not come true.

Martin had made specific predictions. One, delivered via trance by an alien called "the Creator," said a "spaceman" would arrive at midnight on December 20 to rescue Martin and her followers in a "flying saucer." The group made preparations, including learning passwords, cutting the zippers out of their pants, and removing their bras. Festinger, Schachter, and Riecken's book about the experience, *When Prophecy Fails,* describes what happened when the spaceman failed to appear:

> The group began reexamining the original message which had stated that at midnight they would be put into parked cars and taken to the saucer. The first attempt at reinterpretation came quickly. One member pointed out that the message must be symbolic, because it said they were to be put into parked cars but parked cars do not move and hence could not take the group anywhere. The Creator then announced that the message was indeed symbolic, but the "parked cars" referred to their own physical bodies, which had obviously been there at midnight. The flying saucer, he went on, symbolized the inner light each member of the group had. So eager was the group for an explanation of any kind that many actually began to accept this one.

The aliens' big prediction had been the end of the world. But Martin received a new message from the aliens shortly before this was due to occur: "From the mouth of death have ye been delivered. Not since the beginning of time upon this Earth has there been such a force of Good and light as now floods this room."

The group had saved the world! The cataclysm was canceled. Members started calling newspapers to announce the news. They did not even consider the possibility that Martin's prophecies were false.

One of the undercover psychiatrists, Leon Festinger, named this gap between certainty and reality "dissonance." When what we know contradicts what we believe, we can either change our beliefs to fit the facts or change the facts to fit our beliefs. People suffering from certainty are more likely to change the facts than their beliefs.

Next, Festinger studied dissonance in ordinary people. In one experiment, he gave volunteers a mundane task, then asked what they thought of it. Each one said it was boring. Despite this, he persuaded them to tell the next volunteer to arrive that it was fun. After people told someone else the task was fun, their memory of it altered. They "remembered" thinking that the task was fun. They changed what they knew to fit something they had initially only pretended to believe.

Once we become certain, we need the world to become and remain consistent with our certainty. We see things that do not exist and ignore things that do in order to keep life in line with our beliefs. Festinger writes in his 1957 book, *A Theory of Cognitive Dissonance:* "When dissonance is present, in addition to trying to reduce it, the person will actively avoid situations and information which would likely increase the dissonance."

Knowing that dissonance exists does not help prevent it. We can have dissonance about our dissonance. Dorothy Martin had a long career communicating with aliens after her prophecy failed and even after the study about her was published. Some of Martin's followers interpreted the psychologists' research as *proof* of her powers. For example, "Natalina," an "explorer of the supernatural" from Tulsa, Oklahoma, wrote on her website "Extreme Intelligence": "The psychologists determined that when people have a strong enough faith in something, they will often do exactly the opposite of what we would expect when their faith is tested."

How can we know we are seeing something real, and not being deluded by dissonance—that we are like Robin Warren and Judah Folkman, not Percival Lowell and Dorothy Martin?

That's easy: delusion comforts when truth hurts. When you feel sure, feel wary. You may be suffering from certainty.

Delusion's comfort comes from certainty. Certainty is the low road past questions and problems. Certainty is cowardice—the flight from the possibility that we might be wrong. If we already know we are right, why confront queries or qualms? Just climb the Eiffel Tower and fly already.

Confidence is a cycle, not a steady state, a muscle that must be strengthened daily, a feeling we renew and increase by enduring the adversity of creation. Certainty is constant. Confidence comes and goes.

Make an enemy of certainty and befriend doubt. When you can change your mind, you can change anything.

CHAPTER 5

WHERE CREDIT IS DUE

1 | ROSALIND

Sleet like crystal tears fell on cobbles of black umbrellas at the United Jewish Cemetery in London. It was April 17, 1958. Across the sea in Brussels, the World's Fair opened with a scale model of a virus as the main attraction. In the cemetery, a casket containing the body of the scientist who built the model was put in the ground. Her name was Rosalind Franklin. She had died of cancer the day before, aged thirty-seven. Her work was understanding the mechanics of life.

With its gas chambers, guided missiles, and fission bombs, World War II was an apex for the engineering of death. After the war, scientists sought a new summit. Physicist Erwin Schrödinger captured the spirit of the age with a series of talks in Dublin called "What Is Life?" He said the laws of physics are based on entropy: the "tendency of matter to go over into disorder." Yet life resists entropy, "avoiding the rapid decay into the inert," by means then enigmatic. Schrödinger set a bold goal for the science of the rest of the century: to discover how life lives.

Of all the things in the universe, only life escapes inertia and decay, however briefly. An individual organism delays destruction by consuming matter from the environment—by breathing, eating, and drinking, for example—and using it to replenish itself. A species delays destruction by transferring its blueprint from parent to child. Life itself delays destruction by adapting and diversifying these blueprints. At the start of the 1950s, life's mechanism was a mystery; by the end of the decade, much of the mystery had been solved. Rosalind Franklin's model of a virus at the World's Fair was a celebration of that triumph.

The model showed the tobacco mosaic virus, known to scientists as "TMV" and studied throughout the world because it is easy to obtain, highly infectious, and relatively simple. TMV was named for the destruction it wreaks on tobacco leaves, which it stains in a brown patchwork like a mosaic. In 1898, Dutch botanist Martinus Beijerinck showed that the infection was not caused by bacteria, which are relatively large and cellular, but by something smaller and cell-less. He called it a "virus," using the Latin word for "poison."

Bacteria are cells that divide to reproduce, like the cells in other life-forms. A virus has no cells. It occupies, or infects, cells and repurposes their engines of reproduction to make copies of itself—it is a microbiological cuckoo. A virus contains the information it needs to make a copy of itself and little else. But how is the information stored? How does the virus duplicate the information in a new cell without giving away its only copy?

The questions were more important than tobacco or viruses. *All* reproduction is like viral reproduction. Parents do not cut themselves in half to make a child. Like viruses, fathers provide information only: a sperm is a message wrapped in matter. To understand a virus is to understand life.

Life's information is a series of instructions that give cells particular functions. A child is not, as scientists of the nineteenth century believed, a blend of its parents; it inherits discrete instructions from each parent. These discrete instructions are called "genes."

Genes were discovered in 1865 by Gregor Mendel, a friar at St. Thomas's Abbey in Brünn, now part of the Czech Republic. Mendel grew, cross-fertilized, and analyzed tens of thousands of pea plants and found that traits present in one plant could be introduced into its offspring but, in most cases, could not be blended. For example, peas could either be round or wrinkled but could not be both; nor could they be of some intermediate form. When Mendel crossed round and wrinkled peas, their descendants were always round and never wrinkled. The instruction "Be round" dominated the instruction "Be wrinkled." Mendel called these instructions "characters"; today we call them "genes."

Mendel's work was ignored—even Darwin was unaware of it—until 1902, when it was rediscovered and became the basis of "chromosome theory." Chromosomes are packets of protein and acid found in the nuclei of living cells. Their name comes from one of the first things discovered about them—that they become brightly colored when stained during scientific experiments: *chroma* is Greek for "color," and *soma* is Greek for "body." Chromosome theory, developed in parallel by Walter Sutton and Theophilus Painter and formalized by Edmund Beecher Wilson, explained what chromosomes do: they carry the genes that enable life to reproduce.

At first, scientists assumed the chromosome's proteins were the source of life's information. Proteins are long, complicated molecules. Acids, the other component of chromosomes, are relatively simple.

Rosalind Franklin believed life's messengers might be the chromosome's acids, not its proteins. She came to the subject indirectly. During her college years, she developed an interest in crystals and learned how to study them using X-rays. She became an expert on the structure of coal—or, as she called it, "holes in coal"—which gave her a reputation as a talented X-ray crystallographer. This led her to two research positions at the University of London, where she analyzed biological samples instead of geological samples. It was during her second appointment, at Birkbeck College, that she studied the tobacco mosaic virus.

The word "crystal" evokes brittle things like snowflakes, diamonds, and salt, but in science, a crystal is any solid with atoms or molecules arranged in a three-dimensional repeating pattern. Both of the acids found in chromosomes, deoxyribonucleic acid and ribonucleic acid, or DNA and RNA, are crystals.

Crystal molecules are tightly packed: the gap between them is a few ten-billionths of a meter long. Light waves are hundreds of times longer than this, so light cannot be used to analyze a crystal's structure—it cannot pass through the crystal's gaps. But the waves of an X-ray are the same size as the crystal's gaps. They can pass through the crystal's lattice, and as they do so, they are diverted (or "refracted") every time they hit one of the crystal's atoms. X-ray crystallographers deduce the structure of a crystal by sending X-rays through it from every possible angle, then analyzing the results. The work requires precision, attention to detail, and an ability to imagine in three dimensions. Franklin was a master crystallographer.

She needed all her skill to solve the problem of how viruses reproduce. Unlike bacteria, viruses are metabolically inert—meaning that they don't change in any way or "do" anything—if they have not penetrated a cell. The tobacco mosaic virus, for example, is just a tube of motionless protein molecules until it infects a plant—a tube that contains deadly instructions, encoded in RNA. By the time Franklin took on the problem, it had already been determined that there was nothing but empty space in the center of the tobacco mosaic virus. So where were the deadly instructions?

The answer, she discovered, was that the virus is structured like a drill bit: its protein exterior is twisted with grooves, and its core is scored with spirals of acid. This weaponlike form also shows how viruses work. The protein punctures the cell, and then the RNA unspools and takes over the reproductive machinery in the nucleus of the cell, cloning itself and so spreading infection.

Franklin published her results at the start of 1958. She did the work despite being treated for cancer, from which she had been suffering since 1956. The tumors went away, then returned to kill her. She made the model for the World's Fair while she was dying.

Her death was noted in the *New York Times* and the London *Times*. Both newspapers described her as a skilled crystallographer who helped discover the nature of viruses.

Then, in the years after her death, a new truth was told. Rosalind Franklin's contribution to humanity was far greater than her work on the tobacco mosaic virus. For a long time, the only people who knew what she had really accomplished were the three men who had secretly stolen her work: James Watson, Francis Crick, and Maurice Wilkins.

2 | THE WRONG CHROMOSOMES

Watson and Crick were researchers at Cambridge University. Wilkins had been Franklin's colleague and supervisor during her first University of London appointment, at King's College. All three men wanted to be first to answer the question of the age: what is the structure of DNA, the acid that carries the information of life, and how does it work? The men saw themselves in competition. Wilkins called the trio "rats" and wished the other two "happy racing."

Rosalind Franklin was aware of the race but did not compete in it. She believed racing made hasty science, and she had a handicap: she was a woman.

From a genetic perspective, the difference between a man and a woman is one of forty-six chromosomes. Women have two X chromosomes. Men have an X and a Y chromosome. The Y chromosome carries 454 genes, fewer than 1 percent of the total number in a human being. Because of this tiny difference, the creative potential of women has been suppressed for most of human history.

In some ways, Rosalind Franklin was lucky. She was educated at Cambridge University's Newnham College. Had she been born a few generations earlier, she would not have been admitted to Cambridge. Newnham was founded in 1871, the second of the university's women-only colleges. The other, Girton, was founded in 1869. Cambridge University was founded in 1209. For its first 660 years—more than 80 percent of its existence—no women were admitted. Even when they were admitted, women were not equal to men. Despite placing first in the university's entrance exam for chemistry, Franklin could not be a member of the university or an undergraduate. Women were "students of Girton and Newnham Colleges." They could not earn a degree. And even this place in the university's underbelly was a privilege. The number of women allowed to attend Cambridge was capped at five hundred, to ensure that 90 percent of students were men.

Science, while pretending to be dispassionate and rational, has long been an active oppressor of women. Britain's Royal Society of scientists barred women for almost three hundred years, on grounds including the argument that women were not "legal persons." The first women were admitted to the society in 1945. Both were from fields similar to Franklin's: Kathleen Lonsdale was a crystallographer, Marjory Stephenson a microbiologist.

Marie Curie, history's most famous female scientist, did no better. The French Academy of Sciences—the equivalent of Britain's Royal Society—rejected her application for membership. Harvard University refused to award her an honorary degree because, in the words of Charles Eliot, then president emeritus, "Credit does not entirely belong to her." Eliot assumed that her husband, Pierre, did all her work; so did almost all her male peers. They had no such problems assuming that credit "entirely belonged to" any of the men they wanted to honor.

These rejections came despite Curie being the first woman to win a Nobel Prize in science and the only person, male or female, to win Nobel Prizes in two different sciences (for physics in 1903 and chemistry in 1911). The prizes were, in part, a result of her fighting for the credit she deserved. When she accepted her second Nobel Prize, Curie used the word "me" seven times at the start of her speech, stressing, "The chemical work aimed at isolating radium in the state of the pure salt, and at characterizing it as a new element, was carried out specially by me." The second woman to win a Nobel Prize in science was Curie's daughter Irène. Both women shared their prizes with their husbands, except for Marie's chemistry prize, which was awarded after Pierre Curie's death.

The Curies' success did not help Lise Meitner. She discovered nuclear fission only to see her collaborator Otto Hahn receive the 1944 Nobel Prize for her work. The third woman to win a Nobel in science—and the first non-Curie—was biochemist Gerty Cori, in 1947, who, like both the Curies, shared the prize with her husband. The first woman to win without her husband was physicist Maria Goeppert-Mayer, in 1963. In total, only 15 women have won Nobel Prizes in science, compared to 540 men, making women 36 times less likely to win than a man. The odds have changed little since Marie Curie's day: a female scientist wins a Nobel about once every seven years. Only two women other than Curie have won a prize by themselves; there has been only one year, 2009, when women have won prizes in two of the three science categories at the same time; women have never won science prizes in the same category in two consecutive years; and ten of the sixteen prizes given to women have been in the "medicine or physiology category." Only two women not named Curie have won Nobel Prizes in Chemistry. Only one woman not named Curie has won a Nobel Prize in Physics.

This is not because women have less aptitude for science. Rosalind Franklin, for example, took better pictures of DNA than anyone had taken before, then used a complex mathematical equation called the "Patterson function" to analyze them. The equation, developed by Arthur Lindo Patterson in 1935, is a classic technique in X-ray crystallography. The two main properties of electromagnetic waves are their intensity, or "amplitude," and their length, or "phase." The image created by an X-ray shows amplitude but not phase, which can also be a rich source of information. The Patterson function overcomes this limitation by calculating the phase based on the amplitude. In the 1950s, before computers or even calculators, this work took months. Franklin had to use a slide rule, pieces of paper, and hand calculations to work out the phases for every image, each one of which represented a slice of the three-dimensional crystal molecule she was analyzing.

While Franklin was concluding this work, her King's College colleague Maurice Wilkins showed her data and pictures to James Watson and Francis Crick, without her consent or knowledge. Watson and Crick leapt to the conclusion Franklin was diligently proving—that the structure of DNA was a double helix—published it, then shared the Nobel Prize with their secret source, Wilkins. When Rosalind Franklin died, she did not know the three men had stolen her work. Even after she was dead, they did not give her credit. She was not thanked in their Nobel acceptance speech, unlike several men who made lesser contributions. Wilkins referred to Franklin only once in his Nobel lecture and misrepresented her importance by saying that she made "very valuable contributions to the X-ray analysis," rather than confessing that she did *all* the X-ray analysis and far more besides. Watson and Crick did not mention her at all in their Nobel lectures.

3 | THE TRUTH IN CHAINS

Rosalind Franklin was the most important person in the story of DNA's discovery. She was the first-ever member of the human race—or any other species on earth—to see the secret of life. She answered Schrödinger's question "What Is Life?" with a photograph she took on May 1, 1952. She pointed her camera at a single strand of DNA fifteen millimeters, or five-eighths of an inch, from the lens, set the exposure time for one hundred hours, and opened the shutter. It really was *her* camera. She had designed it and overseen its construction in the King's College workshop. It tilted precisely so she could take pictures of DNA specimens at different angles. It was able to take photographs at very close range. It protected DNA specimens from humidity with a brass-and-rubber seal that also allowed Franklin to remove the air around a sample and replace it with hydrogen, a better medium for crystallography. There was nothing else like it anywhere in the world.

Four days later, the picture was ready. It is one of the most important images in history. To any but the most trained eye, it does not look like much: a shadowy circle around something like a ghostly face, its eyes, eyebrows, nostrils, and dimples symmetrically and diagonally aligned, smiling like a Buddha or perhaps God Him- or Herself.

It was clear to Franklin what the picture showed. DNA had the shape of two helixes, like a spiral staircase with no central support. The shape gave a clear indication of how life reproduced. The spiral staircase could copy itself by unwinding and replicating.

Franklin knew what she had, but she did not run through the King's College corridors shouting some equivalent of "Eureka!" She was determined not to leap to conclusions. She wanted to work through the math and have proof before she published, and she was determined to keep an open mind until she had gathered all the data. So she gave the image the serial number 51 and continued her work. She was still completing her Patterson function calculations, and there were many more pictures to take. Then Maurice Wilkins showed picture 51 to James Watson and Francis Crick, and the three men were awarded the Nobel Prize for a woman's work.

It was the same when Marietta Blau, an unpaid woman working at the University of Vienna, developed a technique for photographing atomic particles. Blau could not get a paid position anywhere, even though her work was a major advance in particle physics. C. F. Powell, a man who "adopted and improved" her techniques, was awarded the Nobel Prize in 1950. Agnes Pockels was denied a college education because she was a woman, taught herself science from her brother's textbooks, created a laboratory in her kitchen, and used it to make fundamental discoveries about the chemistry of liquids. Her work was "adopted" by Irving Langmuir, who won a Nobel Prize for it in 1932. There are many similar stories. A lot of men have won Nobel Prizes in science for discoveries made in whole or part by women.

4 | THE HARRIET EFFECT

Even in our new post-genomic age, the game of claims is rigged in favor of white men. One reason is an imbalance first recorded fifty years ago by a sociologist named Harriet Zuckerman. Zuckerman was trying to find out if scientists were more successful alone or in teams. She interviewed forty-one Nobel Prize winners and discovered something that forever changed the direction of her research: that, after winning the prize, many Nobel laureates became wary of joining teams because they find they receive too much individual credit for things the group has done. One said, "The world is peculiar in this matter of how it gives credit. It tends to give the credit to already famous people." Another: "The man who's best known gets more credit, an inordinate amount of credit." Almost every Nobel Prize–winning scientist said the same thing.

Until Zuckerman, most scholars assumed that the strata of science were more or less meritocratic. Zuckerman showed that they are not. More-recognized scientists get more recognition, and less-recognized scientists get less recognition, no matter who does the work.

Zuckerman's discovery is known as the Matthew effect, after Matthew 25:29 — "For whoever has will be given more, and they will have an abundance. Whoever does not have, even what they have will be taken from them." This was the name Robert Merton, a far more eminent sociologist, gave Zuckerman's findings. Zuckerman discovered the effect, then experienced it: the credit for Zuckerman's work went to Merton. Merton gave Zuckerman full acknowledgment, but it made little difference. As she'd predicted, he had recognition and so was given more. There were no hard feelings. Zuckerman collaborated with Merton, then married him.

The Matthew effect—or perhaps more correctly the Harriet effect—is part of the broader problem of seeing what we think, instead of seeing what is. It is unusual that the scientists in Zuckerman's study were honest enough to know they were getting credit they did not deserve. As we are prejudiced about others, so we are prejudiced about ourselves. For centuries, white men have tried to persuade other people that white men are superior. In the process, many white men have become convinced of their own superiority. People often give and take credit based on their prejudices. If there is a person from a "superior" group in the room when something is created, members of the group often assume that the "superior" person did most of the work, even when the opposite is true. Most of the time, the "superior" person makes the same assumption.

I was once forwarded an e-mail that a senior, white, male scientist had sent to a junior, non-white, female scientist. She was applying for a patent. The male scientist demanded to be listed as an inventor on her patent, on the grounds that her research might have been "connected" to him. He claimed he had no interest in getting credit—he was only "making sure she did things correctly." Patent law is complicated, but the patent office's definition of inventorship is not. "Unless a person contributes to the conception of the invention," it reads, "he is not an inventor." If the female scientist named the male scientist as an inventor, she risked invalidating her patent. If she did not, she risked her career. The male scientist's ploy works: he is named as an inventor on nearly fifty patents, an improbable number, especially as most of the patents have many inventors, even though the average number of people who "contribute to the conception" of an invention is two. The man sincerely believed he must have had something to do with the woman's invention, even though the first time he heard of it was when he saw her patent application.

5 | SHOULDERS, NOT GIANTS

Harriet Zuckerman's husband, Robert Merton, was a magnet for credit, and not just because he was a man—he was also one of the most important thinkers of the twentieth century. Merton founded a field called the "sociology of science," which, along with his friend Thomas Kuhn's "philosophy of science," scrutinizes the social aspects of discovery and creation.

Merton dedicated his life to understanding how people create, especially in science. Science claims to be objective and rational, and while its results sometimes are, Merton suspected that its practitioners are not. They are people, capable of being as subjective, emotional, and biased as everybody else. This is why "scientists" have been able to justify so many wrong things, from racial and gender inferiority to canals on Mars and the idea that the body is made of "humors." Scientists, like all creative people, operate in environments—Merton divided them into what he called microenvironments and macroenvironments—which shape what they think and do. The way of seeing that Kuhn called a "paradigm" is part of the macroenvironment; whose contributions are recognized and why is part of the microenvironment.

One of Merton's observations was that the very idea of giving sole credit to any individual is fundamentally flawed. Every creator is surrounded by others in both space and time. There are creators working alongside them, creators working across the hall from them, creators working across the continent from them, and creators long dead or retired who worked before them. Every creator inherits concepts, contexts, tools, methods, data, laws, principles, and models from thousands of other people, dead and alive. Some of that inheritance is readily apparent; some of it is not. But every creative field is a vast community of connection. No creator deserves too much credit because every creator is in so much debt.

In 1676, Isaac Newton described this problem when he wrote, "If I have seen further it is by standing on the shoulders of giants." This may seem like modesty, but Newton used it in a letter where he was arguing with rival scientist Robert Hooke about credit. The comment became famous, and Newton is frequently cited as if he coined the phrase. But Newton was already standing on the shoulders of another when he wrote that sentence. Newton got it from George Herbert, who in 1651 wrote, "A dwarf on a giant's shoulders sees farther of the two." Herbert got it from Robert Burton, who in 1621 wrote, "A dwarf standing on the shoulders of a giant may see farther than a giant himself." Burton got it from a Spanish theologian, Diego de Estella, also known as Didacus Stella, who probably got it from John of Salisbury, 1159: "We are like dwarfs on the shoulders of giants, so that we can see more than they, and things at a greater distance, not by virtue of any sharpness of sight on our part, or any physical distinction, but because we are carried high and raised up by their giant size." John of Salisbury got it from Bernard of Chartres, 1130: "We are like dwarfs standing upon the shoulders of giants, and so able to see more and see farther than the ancients." We do not know from whom Bernard of Chartres got it.

Robert Merton pieced this chain of custody together in a book, *On the Shoulders of Giants,* to exemplify the long, many-handed sequence of gradual improvement that is creation's reality and to show how one person, usually famous, can accumulate credit they do not deserve. Newton's line was, in fact, close to a cliché at the time he wrote it. He was not pretending to be original; it was such a common aphorism that he did not need to cite a source. His reader, Hooke, would have already been familiar with the idea.

But there is a problem with the statement, whether we attribute it to Newton or somebody else: the idea of "giants." If everybody sees further because they are standing on the shoulders of giants, then there are no giants, just a tower of people, each one standing on the shoulders of another. Giants, like geniuses, are a myth.

How many people are holding us up? A human generation is about twenty-five years long. If it was not until fifty thousand years ago that our transition to *Homo sapiens sapiens* — creative people — was complete, then everything we make is built upon two thousand generations of human ingenuity. We do not see further because of giants. We see further because of generations.

6 | INHERITANCE

Rosalind Franklin, master crystallographer, stood on a tower of generations when she became the first person to see the secret of life.

Almost nothing was known about crystals at the start of the twentieth century, but they had been a subject of curiosity at least since the winter of 1610, when Johannes Kepler wondered why snowflakes had six corners. Kepler wrote a book, *The Six-Cornered Snowflake,* in which he speculated that solving the riddle of the snowflake, or "snow crystal," would allow us to "recreate the entire universe."

Many people tried to understand snowflakes, including Robert Hooke, the recipient of Newton's "shoulders of giants" letter. They were drawn, described, and categorized for three centuries, but never explained. No one understood what a snowflake was, because no one understood what a crystal was, because no one understood the physics of solid matter.

The crystals' mysteries are invisible to the eye. To see them, Rosalind Franklin needed a tool that also has its origins in Kepler's time: the X-ray.

While Kepler's curiosity about snowflakes has a clear connection to crystals, the origin of the X-ray starts with something less obvious: improvements in air-pump technology that enabled scientists to wonder about vacuums. One such scientist was Robert Boyle, who used vacuums to try to understand electricity. Others improved on Boyle's work until, almost two hundred years later, German glassblower Heinrich Geissler created the "Geissler tube," a partial vacuum in a bottle that glowed with light whenever an electrical coil connected to it was discharged. Geissler's invention was a novelty, an "interesting scientific toy," during his lifetime, but decades later, it became the basis for neon lighting, incandescent lightbulbs, and the "vacuum tube"—the principal component of early radios, televisions, and computers.

In 1869, English physicist William Crookes built on Geissler's work to create the "Crookes tube," which had a better vacuum than the Geissler tube. The Crookes tube led to the discovery of cathode rays, later renamed "electron beams."

Then, in 1895, German physicist Wilhelm Röntgen noticed a strange shimmering in the dark while he was working with a Crookes tube. He ate and slept in his lab for six weeks while he investigated, then one day positioned his wife's hand on a photographic plate and pointed his Crookes tube at it. When he showed her the result, a picture of her bones, the first ever image of a living skeleton, she said, "I have seen my death." Röntgen named his discovery after the symbol for something unknown: "X-ray."

But what were these unknown rays? Were they particles, like electrons, or waves, like light?

Physicist Max von Laue answered this question in 1912. Laue put crystals between X-rays and photographic plates and found that the X-rays left interference patterns—which are similar to sunlight reflecting off rippling water—on the plates. Particles could not fit through the densely packed molecules of a crystal—and, if they did, they were unlikely to make interference patterns. Therefore, Laue concluded, X-rays were waves.

Within months of hearing about Laue's work, a young physicist named William Bragg showed that the interference patterns also revealed the inner structure of the crystal. In 1915, at the age of twenty-five, he won the Nobel Prize in Physics for his discovery, becoming the youngest ever Nobel laureate. His father, also a physicist called William, received the award too, but this was all "Matthew effect." Bragg the elder played almost no role in his son's discovery.

Bragg's work transformed the study of crystals. Before Bragg, crystallography was a branch of mineralogy, part of the science of mines and mining, and much of the work involved collecting and cataloging; after Bragg, the field became "X-ray crystallography," a wild frontier of physics inhabited by scientists intent on penetrating the mysteries of solid matter.

The sudden shift had an important and unexpected consequence: it advanced the careers of female scientists. In the late 1800s, universities had started admitting women into science classes, albeit reluctantly. Crystallography, a relative backwater, was a field of study where women had been able to find work after graduating. One, a woman named Florence Bascom, was teaching geology at Bryn Mawr College in Pennsylvania, while Bragg was accepting his Nobel Prize. Bascom was the first woman to receive a PhD from Johns Hopkins University, where she was forced to take classes sitting behind a screen so that she would "not distract the men"; she was also the first female geologist appointed by the United States Geological Survey, and had been an expert in crystals long before physicists became interested in them.

When the study of crystals moved from understanding their exterior—mineralogy and chemistry—to understanding their interior—solid-state physics—Bascom followed, taking her female students with her.

One of them was a woman named Polly Porter, who had been forbidden from going to school because her parents did not believe girls should be educated. When Porter was fifteen, her family moved from London to Rome. While her brothers studied, Porter wandered the city, collecting fragments of stone, cataloging the marble the ancient Romans had used to build the capital of their empire. When the family moved to Oxford, Porter found bits of Rome there, too: in Oxford University's Museum of Natural History, which had a collection of ancient Roman marble in need of cleaning and labeling. Henry Miers, Oxford's first professor of mineralogy, noticed Porter's regular visits to the collection and hired her to translate the catalog and reorganize the stones. She discovered crystallography through Miers. He told Porter's parents she should apply for admission to the university, but they would not hear of it.

Porter took a job dusting instead. But not just any dust—the dust in the laboratory of Alfred Tutton, a crystallographer at London's Royal School of Mines. Tutton taught Porter how to make and measure crystals. Then the Porters moved to the United States, so Polly cataloged more stones, first at the Smithsonian Institution, then at Bryn Mawr College, where Florence Bascom discovered her and appealed to Mary Garrett, a suffragist and railroad heiress, for funds so she could study. There she stayed until 1914, the year Bragg's Nobel Prize was announced and crystallography moved from the margins of geology to the foundation of science. At that point, Bascom wrote to Victor Goldschmidt, a mineralogist at Heidelberg University, in Germany:

Dear Professor Goldschmidt:

I have long had the purpose of writing you to interest you in Miss Porter, who is working this year in my laboratory and whom I hope you will welcome in your laboratory next year. Her heart is set upon the study of crystallography and she should go to the fountainhead of inspiration.

Miss Porter's life has been unusual, for she has never been to school or college. There are therefore great gaps in her education, particularly in chemistry and mathematics, but to offset this I believe you will find that she has an unusual aptitude and an intense love of your subject. I want to see her have the opportunities which have so long been denied her. I am both ambitious for her and with faith in her ultimate success.

Yours truly,
Florence Bascom

Goldschmidt welcomed Porter in June 1914.

The next month the First World War started.

Porter succeeded at her work of learning the art of crystallography despite the difficulties of the war and the depression and distraction of Goldschmidt, and three years later, she earned a science degree from Oxford. She stayed at Oxford, conducting research into, and teaching undergraduates about, the crystals that were her passion until she retired, in 1959. One of her most enduring acts was to inspire and encourage a woman who would become one of the world's greatest crystallographers and Rosalind Franklin's mentor: Dorothy Hodgkin.

Hodgkin was a child at the dawn of the crystal revolution. She was two years old when Bragg invented X-ray crystallography, five years old when he and his father won the Nobel Prize, and when she was fifteen, she listened as the elder Bragg gave the Royal Institution's Christmas Lectures for children. In Britain, the lectures, which were started by Michael Faraday in 1825, are as much part of the season as feasting and caroling. Bragg's topic in 1923 was "The Nature of Things"—six lectures describing the recently revealed subatomic world.

"In the last twenty-five years," he noted, "we have been given new eyes. The discoveries of radioactivity and of X-rays have changed the whole situation: which is indeed the reason for the choice of the subject of these lectures. We can now understand so many things that were dim before; and we see a wonderful new world opening out before us, waiting to be explored."

Three of Bragg's lectures were about crystals. He explained their allure: "The crystal has a certain charm due partly to glitter and sparkle, partly to perfect regularity of outline. We feel that some mystery and beauty must underlie the characteristics that please us, and indeed that is the case. Through the crystal we look down into the first structures of nature."

The lectures inspired Dorothy Hodgkin to pursue a career in crystallography, but Oxford disappointed her: crystal structures were a small part of the university's undergraduate science syllabus. It was only in her final year that she met Polly Porter, who was teaching crystallography while also conducting research to classify every crystal in the world. Porter inspired Hodgkin anew and may have even stopped her from straying into another field. Hodgkin wrote, "There was such a mass of material clearly already available on crystal structures that I had not known about—I wondered, for a moment, whether there was anything for me to find out—and gradually realized the limitations of the present which we could pass."

What Hodgkin saw before most other scientists was that X-ray crystallography could be applied not only to rocks but also to living molecules and that it might be able to reveal the secrets of life itself. In 1934, shortly after graduating, she set about proving her idea by analyzing a crystalline human hormone: insulin. The molecule would not yield to the technology of the 1930s. In 1945, she determined the crystal structure of a form of cholesterol, the first ever biomolecular structure to be identified, or "solved," then determined the structure of a second biomolecule, penicillin. In 1954, she worked out the structure of vitamin B_{12}, and for this discovery she was awarded the Nobel Prize.

That same year, Japanese physicist Ukichiro Nakaya solved the mystery of the snowflake. Snowflakes that form at temperatures higher than -40 degrees Celsius are not pure water. They form around another particle, almost always biological, and usually a bacterium. It is a beautiful coincidence that life, in the form of a bacterium, is the nucleus of an abundant crystal, snow, and that a crystal, DNA, is the nucleus of abundant life. Nakaya also showed why snowflakes have six corners: because snowflakes grow from ice crystals, and the crystalline structure of ice is hexagonal.

When Rosalind Franklin started analyzing DNA using X-ray crystallography, she was inheriting a technique pioneered by Dorothy Hodgkin, who was inspired by Polly Porter, who was a protégée of Florence Bascom, who broke ground for all women in science, following work by William Bragg, who was inspired by Max von Laue, who followed Wilhelm Röntgen, who followed William Crookes, who followed Heinrich Geissler, who followed Robert Boyle.

Even the greatest individual contribution is a tiny step on humanity's way. We owe nearly everything to others. Generations are also generators. The point of the fruit is the tree, and the point of the tree is the fruit.

Today, the whole world stands on Rosalind Franklin's shoulders. Everybody benefits from her work; it is a link in the long chain that led to—among many other things—virology, stem-cell research, gene therapy, and DNA-based criminal evidence. Franklin's impact, along with Bragg's, Röntgen's, and all of the others', has even traveled beyond this planet. NASA's robotic rover *Curiosity* analyzes the surface of Mars using onboard X-ray crystallography. Nucleobases, essential components of DNA, have been found in meteorites, and glycolaldehyde, a sugarlike molecule that is a part of RNA, has been discovered orbiting a star four hundred million light-years away from us. Because we have found these buildings blocks so far away, it now seems possible that life is not rare but everywhere. Life was mysterious when Franklin first photographed it; today, we understand it so well we can reasonably suspect that the universe may be full of it.

Rosalind Franklin died because of her DNA. She was an Ashkenazi Jew, descended in part from people who migrated from the Middle East to the shores of Europe's Rhine River during the Middle Ages. Her family name was once Fraenkel; her ancestors were from Wrocław, now in Poland, then the capital of Silesia. Much of her genetic inheritance was European, not Asian: the Ashkenazim began when Jewish men converted European women and survived by prohibiting marriage outside their group. Three of these people had genetic flaws: two of them had mutated breast cancer type 1 tumor suppressor genes, called BRCA1 genes; another had a mutation called 6174delT in his or her breast cancer type 2 tumor suppressor, or BRCA2, gene. Franklin likely inherited one of these mutated genes. The BRCA2 mutation makes a woman fifteen times more likely to get ovarian cancer; the BRCA1 mutation increases her odds by a factor of thirty. Rosalind Franklin died of ovarian cancer.

None of this could have been imagined before she photographed DNA. Today, Ashkenazi Jewish women, all literal cousins of Rosalind Franklin, can get a test to see if they have the BRCA1 or BRCA2 mutations and take preventative measures if they do. These measures are crude: they include surgical removal of both breasts, to reduce the risk of breast cancer, and surgical removal of the ovaries and fallopian tubes, to reduce the risk of ovarian cancer. But in the near future there will likely be a targeted therapy that prevents the mutation from causing cancer, without the need for surgery. This will also be true of other genetic mutations, other cancers, and other diseases. Franklin could not save her own life, but she could and did help save the lives of tens of thousands of other women who were born after her death, many of whom will never know her name.

None of this would have happened, or it would have happened later, if women were still barred from science—not because they are women but because they are human and, thus, as likely to create, invent, or discover as anybody else. The same is true of people who are black, brown, or gay. A species that survives by creating must not limit who can create. More creators means more creations. Equality brings justice to some and wealth to all.

CHAPTER 6

CHAINS OF CONSEQUENCE

1 | WILLIAM

William Cartwright's dog started barking soon after midnight on Sunday, April 12, 1812. There was a single gunshot from the north, one from the south, then one each from the east and the west. Cartwright's watchers awoke at the sounds. Men, unseen and uncounted, came through the night and beat the watchers to the ground in the lee of Cartwright's mill.

Other men broke the mill's windows and pounded on its door with great sledgehammers called "Enochs." Yet more fired pistols through the broken windows, and muskets at the higher floors.

Cartwright, accompanied by five employees and five soldiers, counterattacked, firing muskets from behind raised flagstones and ringing a bell to alert the cavalry stationed one mile away.

The mill door, which Cartwright had reinforced and studded with iron, would not yield to the Enochs. Musket balls smoked up and down. Soon, two men lay dying in the yard. After twenty minutes and

140 shots, the attackers retreated, carrying the wounded, unable to retrieve the dying.

Once the shadows of the mob had disappeared, Cartwright looked out. Hammers and pistols had destroyed his first-floor windows, pane and frame; musket balls had shattered fifty more panes upstairs. His door had been sledged beyond repair. Beyond, two mortally wounded men furled and unfurled among discarded hammers and hatchets, axes, puddles of blood, strips of flesh, and a severed finger.

The object of the attack was Cartwright's automatic loom. The attackers were weavers, trying to destroy the new machine before it destroyed their jobs. They called themselves "Luddites" and had launched similar attacks throughout the north of England. William Cartwright was the first man to ever defeat them.

The Luddites—their name came from the then-famous, possibly fictional machine breaker Ned Ludd—have become icons of both restraint in the face of new technology and entrenched fear of change. They were driven by neither: they were just men desperate to keep their jobs. Their battle was against capital, not technology. The new and improved Enoch sledgehammers they used to wreck looms were named after their inventor, Enoch Taylor, who had also invented the looms that were being wrecked—an irony that was not lost on the Luddites, who chanted, "Enoch did make them, Enoch shall break them."

The Luddites' story is a tale not about right and wrong but about the nuance of new. As our creations advance from generation to generation, they have consequences that, good or bad, are nearly always unforeseen and unintended.

New technology is often called "revolutionary." This is not always hyperbole. The context of that bloody night in England was a collision between two revolutions, one technological and one social.

In the decades before, Europe's monarchs and aristocrats had been besieged. In 1776, thirteen North American colonies had declared independence from King George III of England. The French Revolution started in 1789, and the French king Louis XVI was dead within four years. Thomas Paine summarized the spirit of revolution, and the age, in 1791, when he wrote in *The Rights of Man,* "Governments must have arisen either out of the people or over the people."

At the time of the Luddites, the British government, like the French government that had just been deposed, was one that had arisen *over* the people. The head of state, King George III, was one strand in a cobweb of intermarried, interrelated monarchs covering Europe. George ruled Britain through a tier of intermediaries: hereditary aristocrats who in turn ruled the general population. Recently, a new layer in the social hierarchy had endangered this arrangement: capitalists— men who became wealthy through working and creating work for others, not by accidents of birth. People claiming to be "royal" did not impress the capitalists, who expected political power along with their profits. Their rise was in part a result of inventions like the printing press, which freed information, and labor-saving machines, which freed time. The middle class is a consequence of the creations of the Middle Ages.

The battle at William Cartwright's mill exemplified the new tensions. Cartwright, given but a few of the monarch's soldiers, rang his bell for more, and they never came. The aristocracy was ambivalent about this new industrial class. Many of them recognized the same risk the Luddites saw—that mechanization could concentrate power and wealth in new hands. Technology like Taylor's automatic loom did not threaten one social class. It threatened two.

The Luddites, monarchs, and aristocrats did not fear technology in general so much as the possible consequences of particular technologies for them personally. New tools make new societies.

While the aristocrats were unsure of the threat, the Luddites were certain—so convinced that automatic looms would do them harm that they were willing to risk death, either in their raids or from execution after capture, to stop the rise of the machines. But the longer-term consequences of the looms, a precursor to both computers and robots, were unforeseen, especially by the Luddites. They could never have predicted that their descendants—today's workers—would use information technology and automation to make their living, just as William Cartwright did. In the end, we'll see, it was the working class that gained the most from the new technology. The aristocrats, the only ones who perhaps had the power to keep automation away, did nothing and lost everything.

2 | HUMANITY'S CHOIR

The consequences of technology are mostly unforeseeable, in part because technology is so complex. To understand that complexity, let's step back from Cartwright's mill to consider something apparently all-American and seemingly mundane: a can of Coca-Cola.

The H-E-B grocery store a mile from my home in Austin, Texas, sells twelve cans of Coca-Cola for $4.49.

Each one of those cans originated in a small town of four thousand people on the Murray River in Western Australia called Pinjarra—the site of the world's largest bauxite mine. Bauxite is surface-mined—basically scraped and dug from the top of the ground—and then crushed and washed with hot sodium hydroxide until it separates into aluminum hydroxide and a waste material called "red mud." The aluminum hydroxide is first cooled and then heated to over a thousand degrees Celsius in a kiln, where it becomes aluminum oxide, or alumina. The alumina is dissolved in a molten substance called cryolite, a rare mineral first discovered in Greenland, and turned into pure aluminum using electricity in a process called electrolysis. The pure aluminum sinks to the bottom of the molten cryolite, is drained off, and is placed in a mold. The result is a long, cylindrical bar of aluminum.

Australia's role in the process ends here. The bar is transported west to the port of Bunbury and loaded onto a container ship to begin a month-long journey to—in the case of Coke for sale in Austin—the port of Corpus Christi, on the Texan coast.

After the aluminum bar makes landfall, a truck takes it north on Interstates 37 and 35 to a bottling plant on Burnet Road in Austin, where it is rolled flat in a rolling mill and turned into aluminum sheets. The sheets are punched into circles and shaped into a cup by a mechanical process called drawing and ironing—this not only forms the can but also thins the aluminum. The transition from circle to cylinder takes about a fifth of a second. The outside of the can is decorated using a base layer called "urethane acrylate," then up to seven layers of colored acrylic paint and varnish, which are cured using ultraviolet light. The inside of the can is painted, too—with a chemical called a "comestible polymeric coating," to prevent aluminum from getting into the soda. So far, this vast tool chain has produced only an empty can with no lid. The next step is to fill it up.

Coca-Cola is made from syrup produced by the Coca-Cola Company of Atlanta, Georgia. The syrup is the only thing the Coca-Cola Company provides; the bottling operation belongs to a separate, independent corporation called the Coca-Cola Bottling Company. The main ingredient in the syrup used in the United States is a sweetener called high-fructose corn syrup 55, so named because it is 55 percent fructose, or "fruit sugar," and 42 percent glucose, or "simple sugar"— the same ratio of fructose to glucose as in natural honey. High-fructose corn syrup is made by grinding wet corn until it becomes cornstarch, mixing the cornstarch with an enzyme secreted by a bacillus, a rod-shaped bacterium, and another enzyme, this one secreted by an aspergillus mold, and then using a third enzyme, xylose isomerase, derived from a bacterium called *Streptomyces rubiginosus,* to turn some of the glucose into fructose.

The second ingredient, caramel coloring, gives the drink its distinctive dark brown color. There are four types of caramel coloring; Coca-Cola uses type E150d, which is made by heating sugars with sulfite and ammonia to create bitter brown liquid. The syrup's other principal ingredient is phosphoric acid, which adds acidity and is made by diluting burnt phosphorus (created by heating phosphate rock in an arc furnace) and processing it to remove arsenic.

High-fructose corn syrup and caramel coloring make up most of the syrup, but all they add is sweetness and color. Flavors make up a much smaller proportion of the mixture. These include vanilla, which—as we have already seen—is the fruit of a Mexican orchid that has been dried and cured; cinnamon, which is the inner bark of a Sri Lankan tree; coca leaf, which comes from South America and is processed in a unique U.S. government–authorized factory in New Jersey to remove its addictive stimulant, cocaine; and kola nut, a red nut found on a tree that grows in the African rain forest (this may be the origin of Coca-Cola's distinctive red logo).

The final ingredient, caffeine, is a stimulating alkaloid that can be derived from the kola nut, coffee beans, and other sources.

All these ingredients are combined and boiled down to a concentrate, which is transported from the Coca-Cola Company factory in Atlanta to the Coca-Cola Bottling Company factory in Austin, where it is diluted with local water infused with carbon dioxide. Some of the carbon dioxide turns to gas in the water, and these gas bubbles give the water effervescence, also known as "fizz," after its sound. The final mixture is poured into cans, which still need lids.

The top of the can is carefully engineered: it is aluminum, too, but it has to be thicker and stronger than the rest of the can to withstand the pressure of the carbon dioxide gas, and so it is made from an alloy with more magnesium. The lid is punched and scored, and a tab opening, also made of aluminum, is installed. The finished lid is put on top of the filled can, and the edges of the can are folded over it and welded shut. Twelve of these cans are packaged into a paperboard box called a fridge pack, using a machine capable of producing three hundred such packs a minute.

The finished box is transported by road to my local H-E-B grocery store, where—finally—it can be bought, taken home, chilled, and consumed. This chain, which spans bauxite bulldozers, refrigerators, urethane, bacteria, and cocaine, and touches every continent on the planet except Antarctica, produces seventy million cans of Coca-Cola each day, one of which can be purchased for about a dollar on some close-by street corner, and each of which contains far more than something to drink. Like every other creation, a can of Coke is a product of our world entire and contains inventions that trace all the way back to the origins of our species.

The number of individuals who know how to make a can of Coke is zero. The number of individual nations that could produce a can of Coke is zero. This famously American product is not American at all. Invention and creation, as we have seen, is something we are all in together. Modern tool chains are so long and complex that they bind us into one people and one planet. They are chains of minds: local and foreign, ancient and modern, living and dead—the result of disparate invention and intelligence distributed over time and space. Coca-Cola did not teach the world to sing, no matter what its commercials suggest, yet every can contains humanity's choir.

The story of Coca-Cola is typical. Everything we make depends on tens of thousands of people and two thousand generations of ancestors.

In 1929, Russian Ilya Ehrenburg described how a car was made, much as I have done here for Coca-Cola, in a book called *The Life of the Automobile*. He begins with Frenchman Philippe Lebon developing the first internal combustion engine at the end of the eighteenth century and ends with the emergence of the oil industry. On the way, Ehrenburg shows contributions from, among others, Francis Bacon, Paul

Cézanne, and Benito Mussolini. He writes of Henry Ford's conveyor belts—"It's not even a belt. It's a chain. It's a miracle of technology, a victory of human intelligence, a growth of dividends."

In 1958, Leonard Read traced the history of a yellow "Mongol 482" pencil, made by the Eberhard Faber Pencil Company, from the growth and logging of a cedar tree in Oregon, through its transportation to a milling and painting factory in San Leandro, California, and onward to Wilkes-Barre, Pennsylvania, where it is grooved, laid with lead made from Sri Lankan graphite and Mississippian mud, lacquered with the refined oil of castor beans, and topped with brass and a material called factice, "a rubber-like product made from reacting rape seed oil from the Dutch East Indies with sulphur chloride," to make an eraser.

And, in 1967, Martin Luther King Jr. told a similar story while preaching to the Ebenezer Baptist Church in Atlanta, in a sermon called "Peace on Earth":

> It really boils down to this: that all life is interrelated. We are all caught in an inescapable network of mutuality, tied into a single garment of destiny. Whatever affects one directly, affects all indirectly. We are made to live together because of the interrelated structure of reality. Did you ever stop to think that you can't leave for your job in the morning without being dependent on most of the world? You get up in the morning and go to the bathroom and reach over for the sponge, and that's handed to you by a Pacific islander. You reach for a bar of soap, and that's given to you at the hands of a Frenchman. And then you go into the kitchen to drink your coffee for the morning, and that's poured into your cup by a South American. And maybe you want tea: that's poured into your cup by a Chinese. Or maybe you're desirous of having cocoa for breakfast, and that's poured into your cup by a West African. And then you reach over for your toast, and that's given to you at the

hands of an English-speaking farmer, not to mention the baker. And before you finish eating breakfast in the morning, you've depended on more than half of the world.

Half the world *and* the two thousand generations that came before us. Together, they give us what computer scientists call "tool chains"— the processes, principles, parts, and products that let us create.

King described tool chains to argue for world peace. But the politics and morality of our long and ancient tool chains are complicated. Ilya Ehrenburg described the chain that built the automobile to argue for Marxism: he believed that the industrial processes of mass production endangered and dehumanized workers. Leonard Read saw the pencil's journey as an argument for libertarianism: he claimed that such spontaneous complexity was possible only when people were free of central control from government "masterminds." Clearly we are tempted to ascribe meaning to the complexity of creation. But should we?

3 | AMISH LESSONS

There is a real-life model for exploring the relationship between creation and its consequences: America's Amish people—a group of Mennonite Christians descended from Swiss immigrants. The Amish value small, rural communities, and their way of life includes protecting these communities from external influence. As electrification spread through America during the twentieth century, the Amish resisted it. They did the same with other inventions from the period, notably the car and telephone. As a result, the Amish, particularly traditional or "Old Order" Amish, have a reputation for being old-fashioned, frozen in time, and opposed to technology.

But the Amish do *not* avoid new technology. They are as creative and resourceful as anybody, and more creative and resourceful than most.

They generate electricity with solar panels, have invented sophisticated systems for using batteries and propane gas, use LED lighting, operate machines powered by gas engines or compressed air, make photocopies, refrigerate food, and use computers for word processing and making spreadsheets. The thing they avoid as much as possible is using this technology to connect to the non-Amish, or "English," world. This is why they generate their own power and do not have their own long-distance transportation—they take taxis to travel beyond the range of their horse-drawn buggies—and why their computers do not have Internet access. They are not practicing self-sufficiency. Most of the tools the Amish use are like Coca-Cola: they contain ideas from across the globe; could not be made without large-scale power plants, water-treatment facilities, oil refineries, and information systems; and cannot be sourced locally. The Amish do not have a puritanical preference for manual labor, either: the line between convenience and efficiency is fine, and while the Amish value work, they do not treasure inefficiency. Amish clothes dryers and word processors do things that the Amish could also do—and have previously done—by hand.

Contrary to their reputation, the Amish are among the most conscious, thoughtful tool users in the world. Amish leader Elmo Stoll explains: "We do not consider modern inventions to be evil. A car or television set is a material thing—made of plastic, wood, or metal. Lifestyle changes are made possible by modern technologies. The connection between the two needs to be examined with care."

The Amish approach to technology only seems arbitrary. The Amish are cautious about technology because they are cautious about how it shapes their communities.

The most unusual thing about the Amish may be that they walk their talk. They are not the only people with objections to creation, change, and technology. Some believe that not all technology is good, therefore most technology is bad; that because technology cannot solve all problems, it cannot solve any problems; and that anyone who thinks technology can do good is a naïve optimist, ignorant of technology's harmful consequences. One example is writer and technology critic Evgeny Morozov, who argues against what he calls "the folly of technological solutionism":

> Not everything that could be fixed should be fixed—even if the latest technologies make the fixes easier, cheaper, and harder to resist. Sometimes imperfect is good enough; sometimes, it's much better than perfect. What worries me most is that, nowadays, the very availability of cheap and diverse digital fixes tells us what needs fixing. It's quite simple: the more fixes we have, the more problems we see.

What's more, he says, technology is

> embedded in a world of complex human practices, where even tiny adjustments to seemingly inconsequential acts might lead to profound changes in our behavior. It might very well be that by optimizing our behavior *locally* . . . we'll end up with suboptimal behavior *globally*. . . . One local problem might be solved—but only by triggering several global problems that we can't recognize at the moment.

Morozov is right. "The more fixes we have, the more problems we see" is a good description of Karl Duncker's problem-solution loops, discussed in chapter 2. Problems lead to solutions, which lead to problems, and—Morozov's second point—because solutions are assembled across the world and inherited by future generations, the problems a solution creates may be felt only far away or in the future. Creating can cause problems unintended, unforeseen, and often unknowable, at least in advance. To illustrate, we return to our can of Coca-Cola.

Once we knelt by a stream to scoop water with bare hands. Now we pull a tab on an aluminum can and drink ingredients we cannot name from places we may not know mixed in ways we do not understand.

Coca-Cola is a branch on our fifty-thousand-year-old tree of new. It is there because water is our most important nutrient. If we do not drink water, we die within five days. If we drink the wrong water, we die of waterborne diseases like cyclosporiasis, microsporidiosis, coenurosis, cholera, and dysentery. Thirst should limit us to places within a day or two's walk of potable water and make migration and exploration dangerous. But the two thousand generations developed tools to make water portable and potable and allow us to live far away from rivers and lakes.

Early technologies for carrying and storing water included skins, hollowed gourds called calabashes, and—eighteen thousand years ago—pottery. Ten thousand years ago we developed wells, which allowed constant access to fresh groundwater. Three thousand years ago, people in China started drinking tea, a step that coincided with drinking boiled water, a practice that—coincidentally—killed disease-bearing microorganisms. The existence of these organisms was not discovered for another twenty-five hundred years, but as the technology of tea spread gradually from China through the Middle East and eventually, around 1600 C.E., to Europe, tea drinkers began to suspect that water was healthier when boiled. Boiling also enabled free-ranging travel, as water found along the way could now be made safe.

The best source of pure water is the spring—nature's equivalent of a well, where groundwater flows up from an aquifer. This water, clean and rich in minerals, has been revered for thousands of years; natural springs are often considered sacred sites of healing. Some spring water is naturally effervescent.

As bottles—first developed by the Phoenicians of the Middle East twenty-five hundred years ago—became more common, it was at last possible to transport sacred water, with its healing purity and high mineral content, from springs to other places. Once bottled and transported, these "mineral waters" could also be flavored.

Some of the earliest flavored waters were Persian *sharbats,* or sherbets, made using crushed fruits, herbs, and flower petals, and first described in Ismail Gorgani's twelfth-century medical encyclopedia, *Zakhireye Khwarazmshahi.* About a hundred years later, people in Britain drank water mixed with fermented dandelions and the roots of the burdock plant, which made it effervescent. Hundreds more years later, similar drinks were made in Asia and the Americas, using parts of a prickly Central American vine called sarsaparilla or the roots of sassafras trees. All of these variants on the theme of sparkling water and drinks made with natural ingredients were thought to have health benefits.

In the late 1770s, chemists began to replicate the properties of springwater and herbal drinks. In Sweden, Torbern Bergman made water effervescent using carbon dioxide. In Britain, Joseph Priestley did the same. Johann Jacob Schweppe, a Swiss German, commercialized Priestley's process and started the Schweppes Company in 1783. The mineral content of springwater was replicated with phosphate and citrus to make drinks called orange or lemon "phosphates," or "acids"; these terms were popularly used for flavored effervescent water in the United States into the twentieth century.

As mineralization and carbonation became common, the healing properties associated with springwater receded in favor of remedies and tonics that contained exotic ingredients, such as the fruit of the African baobab tree and roots supposedly extracted from swamps. Many of these "patent medicines" contained cocaine and opium, which made them effective in treating pain (if nothing else) and also addictive.

One of these medicines, invented by chemist John Pemberton in

Georgia in 1865, was made from ingredients including kola nut and coca leaf, as well as alcohol. Twenty years later, when parts of Georgia banned alcohol consumption, Pemberton made a nonalcoholic version, which he called "Coca-Cola." In 1887, he sold the formula to a drugstore clerk named Asa Candler.

A few years earlier, Louis Pasteur, Robert Koch, and other European scientists had discovered that bacteria caused disease, marking the beginning of the end for remedies and tonics. During the next two decades, medicine became scientific, and also regulated. Harvey Washington Wiley, chief chemist at the United States Department of Agriculture, led a crusade that culminated in the signing of the Pure Food and Drug Act in 1906 and the creation of the government agency that became the U.S. Food and Drug Administration.

In retreat as a medicine, Coca-Cola syrup was mixed with carbonated water in drugstores and sold as a beverage, its health claims softened to ambiguous adjectives such as "refreshing" and "invigorating." At first, the carbonated water was added manually, and the drink was available only at soda fountains. Bottling was such a foreign idea that, in 1899, Candler licensed the U.S. bottling rights, in perpetuity, to two young lawyers for one dollar, because he thought that all the money in cola would come from selling the syrup.

This may seem like an amazing mistake, but in 1899 things weren't so obvious. Glass was not easy to mass-produce, and Candler might have assumed that bottling would be a small business forever. But glass and bottling technologies were improving. In 1870, Englishman Hiram Codd developed a soda bottle that used a marble as a stopper—an ingenious approach that took advantage of the pressure from the carbonation to push the marble up the neck of the bottle to form a seal. Today, these Codd bottles sell at auction for thousands of dollars. As bottle technology improved, Coca-Cola bottling increased. Ten years after Candler sold his bottling rights, there were four hundred Coca-Cola bottling plants in the United States. Coca-Cola, once tied to the soda fountain, had become portable, and it would soon migrate again, from the bottle to the can.

The story of the can begins with Napoleon Bonaparte. Napoleon, having lost more soldiers to malnutrition than to combat, had concluded that "an army marches on its stomach." In 1795, the French revolutionary government offered a twelve-thousand-franc prize to anyone who could invent a way to preserve food and make it portable. Nicolas Appert, a Parisian confectioner, spent fifteen years experimenting and ultimately developed a method of preserving food by sealing it in airtight bottles then placing the bottles in boiling water. As with water for tea, the boiling killed bacteria—in this case, the bacteria that caused food to rot, a phenomenon that would not be understood for another hundred years. Appert sent sealed bottles that included eighteen types of food, ranging from partridge to vegetables, to soldiers at sea, who opened them after four months and found unspoiled, apparently fresh food inside. Appert won the prize, and Napoleon awarded it to him personally.

France's enemy, Britain, viewed Appert's preservation technology as a weapon. Preserved food extended Napoleon's reach. The army that marched on its stomach could now march farther. Britain's response was immediate: inventor Peter Durand improved upon Appert's approach by using cans made of tin instead of bottles. King George III awarded him a patent for his invention. Whereas glass bottles were fragile and difficult to transport, Durand's cans were far more likely to survive the march to war. Canned food quickly became popular among travelers. It helped fuel the voyages of German explorer Otto von Kotzebue and British admiral William Edward Parry, as well as the California Gold Rush—which started in 1848 and saw three hundred thousand people move to California, establishing San Francisco as a major city in the process—and it extended the range of both armies in the American Civil War.

In a coincidence that nods to Napoleon and the origins of canning, Coca-Cola developed the first soda cans during the 1950s to supply American soldiers fighting a distant war in Korea. They were manufactured from tin that had been thickened to contain the pressure of carbonation and coated to prevent chemical reactions, steps that made them heavy and expensive. When cheaper, lighter aluminum cans were invented in 1964, Coca-Cola's bottlers adopted them almost immediately.

Coca-Cola exists because we get thirsty. It exists because water can be dangerous and we cannot all live next to a spring. It exists because people got sick and hoped that herbs and roots and tree bark from far-off places might help them. It exists because we sometimes need to travel—to flee, hunt, go to war, or search for better places and ways. Coca-Cola may look like a luxury, but it exists because of a need for life.

Yet, like all creations, Coca-Cola is flawed by unforeseen, unintended, and often distant consequences. Aluminum begins in bauxite surface mines, which are devastating to their local environment. In 2002, a British mining company, Vedanta, requested approval to mine bauxite in the Niyamgiri Hills of East India, home to an indigenous tribal people called the Dongria Kondh. The plan, which was approved by the Indian government, would have destroyed the tribe's way of life, and also their sacred mountain. The tribespeople led international protests that put a stop to the mine, but it was a close call—and one that, of course, had no impact on equally destructive bauxite mines in Australia, Brazil, Guinea, Jamaica, and more than a dozen other countries around the world.

High-fructose corn syrup has been cited as a cause of rising obesity, especially in the United States. Americans ate 113 pounds of sugar per person in 1966. By 2009, this had risen to 130 pounds per person, an increase that may, in part, be due to the introduction of high-fructose corn syrup, which, because of import tariffs, is much cheaper in America than sugar. The average American consumes around forty pounds of high-fructose corn syrup a year.

Caffeine can be intoxicating and addictive if overused and if taken in excess can cause vomiting or diarrhea, which can result in dehydration—the opposite of drinking. Caffeine in soda is a particular problem for children: they now drink an average of 109 milligrams per day—twice as much as children in the 1980s.

Even though aluminum is easily recycled, many aluminum cans are disposed of in landfill sites, where they take hundreds of years to decompose. The production and distribution of each can adds around half a pound of carbon dioxide to the atmosphere, where it contributes to climate change.

The Coca-Cola Company has been an effective proponent of global trade and has succeeded in manufacturing and selling its product all over the world, a strategy that has caused conflict and concern in many countries, including India, China, Mexico, and Colombia. One issue is water rights: the only local ingredient in Coca-Cola is water, and manufacturing twelve ounces of Coke requires far more than twelve ounces of water because of cleaning, cooling, and other industrial processes. When all the processes in Coca-Cola's tool chain are taken into account, a twelve-ounce can uses more than four thousand ounces, or over thirty gallons, of water. It will always be cheaper and more efficient to drink water than Coke, and this is a problem in areas suffering from water shortages.

So, do better tools always lead to a better life? Does making better things always make things better? How can we be sure that making things better won't make things worse?

These are questions we, like the Amish and Evgeny Morozov, must ask. Sometimes technology's flaws are dangerous—even deadly. Coca-Cola's early competitors, root beer and sarsaparilla, were both made with fermented roots of the sassafras tree, an ingredient that is now banned because it is suspected of causing liver disease and cancer. Glass once contained enough lead to cause lead poisoning, one consequence of which can be gout, a painful inflammation that usually affects the joint of the big toe. Gout was long known as the "rich man's disease" because its sufferers were so often from the upper classes of society—people like King Henry VIII, John Milton, Isaac Newton, and Theodore Roosevelt. Benjamin Franklin went so far as to write an essay titled "Dialogue Between Franklin and the Gout." Dated "Midnight, October 22, 1780," the dialogue recounts a conversation in which Franklin asks his gout to explain what he did "to merit these cruel sufferings." He assumed they were the result of too much food and not enough exercise, and "Madam Gout" chides him for his laziness and gluttony. In fact, the cause of the "rich man's disease," for Franklin and all the others, was lead crystal decanters, which were used by the upper classes for storing and serving port, brandy, and whiskey. "Lead crystal" is not crystal at all but glass with a high lead content. The lead can leach from the glass into the alcohol and cause lead poisoning, which causes gout.

Lead poisoning may also have afflicted the majority of Roman emperors, including Claudius, Caligula, and Nero, who drank wine flavored with syrup made in lead pots. This had consequences far beyond gout. Their lead poisoning was so severe that it probably caused organ, tissue, and brain problems—severe symptoms that affected so many emperors that they likely contributed to the end of the Roman Empire.

As Amish leader Elmo Stoll says, new is neutral, neither good nor bad. As Morozov says, new *things* tend to be good for some people and bad for others, or good now and bad later, or both.

Not convinced? Let's return to William Cartwright's mill.

Understanding the impact of the past on the present is as hard as predicting the impact of the present on the future. They probably did not know it, but the weavers who attacked William Cartwright's automatic loom would not have been weavers at all but for automation. Until the thirteenth century, England's textile industry was centered in the southeast. What moved it north to places like Rawfolds, Yorkshire, the site of Cartwright's mill, was mechanization—specifically the mechanization of the cloth-cleaning process known as "fulling." For millennia, fulling cloth was like treading grapes, accomplished by the stomping of naked feet. To pace themselves and stay synchronized, the fullers, usually women, sang special "fulling songs," slow at the beginning when the cloth was tough, then quickening as it became more supple. The women adjusted the length and tempo of their song to fit the size and type of cloth being fulled. For example, from Scotland, originally sung in Gaelic, with nonsense syllables added here and there as needed:

Come on, my love,
Keep your promise to me,
Take greetings from me,
Over to Harris,
To John Campbell,
My brown-haired sweetheart,
Hunter of goose,
Seal and swan,
Of leaping trout,
Of bellowing deer,
Wet is the night,
Tonight and cold.

In England, the tradition of the fulling song was ended by a technology that revolutionized the world between the first and fifteenth centuries: the watermill.

Watermills were invented two thousand years ago, first spinning horizontally, like Frisbees, then vertically, like cart wheels. By the time the millennium turned, they were everywhere, initially used for grinding grain but soon for fulling cloth—as well as tanning, laundering, sawing, crushing, polishing, pulping, and making "milled" coins.

The new importance of rivers changed the value of land. Sites that could deliver energy to mills were now among the most important places in the world. The work went where the energy was.

During the first millennium, England's textile trade was centered in its southeastern counties, but mechanized fulling machines needed a type of waterpower that was available only in the northwest. The textile industry relocated. By the end of the thirteenth century, England's singers of fulling songs were silent.

This revolution in power sowed the seeds of the Enlightenment: the experience of engineering nature's energy led directly to the development of theoretical physics and the scientific revolution. Newton was probably inspired more by a churning watermill than a falling apple.

By the time William Cartwright was born, at the end of the eighteenth century, textile manufacturing was highly automated and had been for centuries. The difference between Cartwright's new loom and his old mill was that the loom replaced mental as well as manual labor. Fulling by treading is a rote task. People supply little more than kinetic energy from their muscles. This is why cranks, cams, and gears attached to waterwheels replaced manual labor so quickly: fulling is mainly applied power. But weaving is mental as well as manual work. It takes mind, not just muscle, to interpret and understand weaving patterns. As waterpower increased the volume of the textile industry, demand for weavers grew, which created a need for workers with better-trained brains. A system of apprenticeship arose to meet this need: master weavers taught teenage children the skill of textile making. Weaving apprenticeships were a common form of schooling in the days before public education: in 1812, the year of the Luddites' attack on William Cartwright's mill, around one in twenty English teenagers living in or near mill towns became weavers' apprentices. It was these same workers with better-trained brains who started to demand political reform during the late eighteenth and early nineteenth centuries.

The automated loom threatened the weavers because it could "think," too, or at least follow directions. Weaving patterns were fed into it using punched cards that could mimic the mind of the weaver, doing his thinking more quickly and precisely, and making him redundant in the process. It was the first programmable machine—in many ways, the first computer. The Luddites were protesting the start of the information revolution.

At the time, the consequences of this revolution seemed bleak. Men descended from manual laborers had been trained to think because mills had reduced the need for manual labor and increased the need for mental work. Now new looms threatened to reduce the need for them, too—perhaps to eliminate the need for workers almost entirely.

What the Luddites could not foresee was that the opposite would happen. The consequence of William Cartwright's victory was entirely unexpected and unintended. The automated loom did not reduce the need for intelligent labor; it increased it. As simple programmable machines took over simple mental tasks, the manufacturing efficiencies that followed created new jobs in a vast new tool chain—jobs like maintaining, designing, and building ever more sophisticated machines; planning production; accounting for income and expenses; and jobs that, less than a century later, would be called "management." These jobs required workers who could do more than think. They required workers who could *read*.

In 1800, one-third of all Europeans could read, in 1850 one-half of all Europeans could read, and in 1900 almost all Europeans could read. After millennia of illiteracy, everything changed in a century. All of your ancestors were probably illiterate until a few generations ago. Why can you read though they could not? The big reason is automation.

The men who attacked Cartwright's mill in 1812 did not learn to read after they lost their campaign against the automated loom, but their children and grandchildren did. Industrialized nations responded to the need for smarter workers by investing in public education. Between 1840 and 1895, school attendance in these countries grew faster than population.

As automation improved and proliferated during the twentieth century, it both drove and was driven by the continued expansion of education. Every year more children were educated to an ever-increasing level. In 1870, America had 7 million elementary school students, 80,000 secondary school students, and awarded 9,000 college degrees. In 1990, America had 30 million elementary school students, 11 million secondary school students, and awarded 1.5 million college degrees. Relative to population, this is almost the same number of children in elementary school but thirty-five times more children in secondary school and twenty-five times more college graduates. The trend toward higher education continues. The number of Americans earning college degrees almost doubled between 1990 and 2010.

The Luddites did not—and could not—foresee this when they tried to wreck Cartwright's loom. Cartwright could not have foreseen it, either. Every man was for himself; none could have imagined the far better future automation would bring to his grandchildren.

Chains of tools have chains of consequences. As creators, we can anticipate some of these consequences, and if they are bad, we should of course take steps to prevent them, up to and including creating something else instead. What we cannot do is stop creating.

This is where self-described technology "heretics" like Evgeny

Morozov go wrong. The answer to invention's problems is not less invention but more. Invention is an act of infinite and imperfect iteration. New solutions beget new problems, which beget new solutions. This is the cycle of our species. We will always make things better. We will never make them best. We should not expect to anticipate all the consequences of our creations, or even most of them, good *or* bad. We have a different responsibility: to actively seek those consequences out, discover them as soon as possible, and, if they are bad, to do what creators do best: welcome them as new problems to solve.

THE GAS IN YOUR TANK

1 | WOODY

In March 2002, Woody Allen did something he had never done. He flew from New York to Los Angeles, put on a bow tie, and attended the Academy of Motion Picture Arts and Sciences annual awards ceremony, the Oscars. Allen had won three Academy Awards and received seventeen other nominations, including more screenwriting nominations than any other writer, yet he had never attended a ceremony. In 2002, his movie *The Curse of the Jade Scorpion* was nominated for nothing. He went anyway. The audience stood and applauded in welcome. He introduced a montage of movie scenes made in New York and encouraged directors to continue working there even though terrorists had attacked the city months earlier. He said, "For New York City, I'll do anything."

Why does Allen avoid the ceremony? He gives several tongue-in-cheek excuses—the two most common being that there is nearly always a good basketball game on that night and that he has to play clarinet every Monday with the Eddy Davis New Orleans Jazz Band.

Neither reason is real. The real reason, which he explains occasionally, is that he believes the Oscars will diminish the quality of his work.

"The whole concept of awards is silly," he says. "I cannot abide by the judgment of other people, because if you accept it when they say you deserve an award, then you have to accept it when they say you don't."

On another occasion: "I think what you get in awards is favoritism. People can say, 'Oh, my favorite movie was *Annie Hall*,' but the implication is that it's the best movie, and I don't think you can make that judgment except for track and field, where one guy runs and you see that he wins; then it's okay. I won those when I was younger, and those were nice because I knew I deserved them."

Whatever motivates Woody Allen, it is not awards. His example is extreme—almost all other Academy Award–nominated writers, directors, and actors attend the Oscars—but it points to something important. Prizes are not always carrots of creation. Sometimes, they can inhibit and impair.

Motives are never simple. We are motivated by a soup of things, some we are aware of and some we are not. Psychologist R. A. Ochse lists eight motivations for creating: the desire for mastery, immortality, money, recognition, self-esteem; the desire to create beauty, to prove oneself, and to discover underlying order. Some of these rewards are internal, some external.

Harvard psychologist Teresa Amabile studies the connection between motivation and creation. Early in her research, she had a suspicion that internal motivation improves creation but external motivation makes it worse.

The external motivator Woody Allen avoids is the evaluation of others. Poet Sylvia Plath admitted to craving what she called "the world's praise," even though she found it made creating harder: "I want to feel my work good and well taken, which ironically freezes me at my work, corrupts my nunnish labor of work-for-itself-as-its-own-reward."

In one of her studies, Amabile asked ninety-five people to make collages. In order to test the role of outside evaluation on the process of creation, some participants were told, "We have five graduate artists from the Stanford Art Department working with us. They will make a detailed evaluation of your design, noting the good points and criticizing the weaknesses. We will send you a copy of each judge's evaluation of your design." Others were given no information about being evaluated.

In fact, all the collages were evaluated on many dimensions by a panel of experts. Work by people expecting evaluation was significantly less creative than work done by people making collages for their own sake. The people expecting evaluation also reported less interest in doing the work—the internal creative drive Plath called "nunnish labor" had been diminished.

Amabile replicated these results in a second experiment with a new variable—an audience. She divided forty people into four groups. She told the first group it was being evaluated by four art students watching from behind a one-way mirror, the second group that it would be evaluated by art students waiting elsewhere, and the third group that behind the mirror people were waiting to begin a different experiment. She did not mention audiences or evaluations to the fourth group. This was the one that did the work that was the most creative. The next most creative group was the other group that had not been told it would be evaluated, while knowing there were people watching. The group that expected to be evaluated but had no audience came in third. The least creative group by a considerable amount was the group that was being both evaluated *and* judged. The evaluated groups reported more anxiety than the nonevaluated groups. The more anxious they were, the less creative they were.

In her next test, Amabile examined written rather than visual creations. She told people they were participating in a handwriting study. As before, there were four groups, some evaluated and some not, some watched and some not. Amabile gave them twenty minutes to write a poem about joy. Once again a panel of experts judged the poems and ranked them from most to least creative. The results were the same. What's more, nonevaluated subjects reported that they were more satisfied with their poems. Evaluated subjects said that writing them felt like work.

Amabile's research validates Woody Allen's reasons for avoiding the Oscars. Allen also skipped high school classes and dropped out of college. Missing award ceremonies is, in his case, part of a pattern of avoiding the potential destruction of external influence.

Allen works at a small desk in the corner of his New York apartment, writing movie scripts on yellow legal paper using a burgundy red Olympia SM2 portable typewriter that he bought when he was sixteen. He says, "It still works like a tank. Cost me forty dollars, I think. I've written every script, every *New Yorker* piece, everything I've ever done on this typewriter."

He keeps a miniature Swingline stapler, two plum-colored staple removers, and scissors alongside the typewriter, and he literally cuts and pastes—or, rather, staples—his writing from one draft to the next: "I have a lot of scissors here, and these little stapling machines. When I come to a nice part, I cut that part off and staple it on."

The result is a mess: a patchwork of paper, each piece either held together with staples or pocked with the acne of staples removed. And covering the mess, in an eleven-point typeface called Continental Elite, colored in a spectrum of grays and blacks that can only come from metal on ribbon, is the screenplay for a movie that will almost certainly be a hit and may, incidentally, win some of the awards Woody Allen avoids.

In 1977, one of these ragged yellow quilts became the movie *Annie Hall*. Allen thought it was terrible: "When I was finished with it, I didn't like the film at all, and I spoke to United Artists at the time and offered to make a film for them for nothing if they would not put it out.

I just thought to myself, 'At this point in my life, if this is the best I can do, they shouldn't give me money to make movies.'"

United Artists released the film anyway. Allen was wrong to doubt it: *Annie Hall* was a great success. Marjorie Baumgarten of the *Austin Chronicle* wrote, "Its comedy, performances, and insights are all dead-on perfect." Vincent Canby of the *New York Times:* "It puts Woody in the league with the best directors we have." Larry David, cocreator of the TV show *Seinfeld:* "It changed the way comedies were going to be made forever."

Allen's views about awards first became clear when *Annie Hall* was nominated for five Academy Awards and he refused to attend the ceremony. He did not even watch it on television. He recalled, "The next morning I got up, and I get the *New York Times* delivered to my apartment, and I noticed on the front page, on the bottom, it said, '*Annie Hall* wins 4 Academy Awards,' so I thought 'Oh, that's great.'"

Two of the awards, for Best Director and Best Screenplay, went to Allen himself. Unimpressed, he insisted that the phrase "Academy Award Winner" could not appear on advertising for the film anywhere within a hundred miles of New York.

Allen's next film was *Stardust Memories.* It underlined his indifference to praise: "It was my least popular film but it was certainly my own personal favorite."

Woody Allen is not alone in wanting to avoid being distracted by the judgment of others. When T. S. Eliot ascended to the highest peak of praise, the Nobel Prize in Literature, he did not want it. Poet John Berryman congratulated him, saying it was "high time." Eliot replied that it was "rather too soon. The Nobel is a ticket to one's own funeral. No one has ever done anything after he got it." His acceptance speech was modest to the point of evasion:

When I began to think of what I should say, I wished only to express very simply my appreciation, but to do this adequately proved no simple task. Merely to indicate that I was aware of having received the highest international honor that can be bestowed upon a man of letters would be only to say what everyone knows already. To profess my own unworthiness would be to cast doubt upon the wisdom of the Academy; to praise the Academy might suggest that I approved the recognition. May I therefore ask that it be taken for granted, that I experienced, on learning of this award to myself, all the normal emotions of exaltation and vanity that any human being might be expected to feel at such a moment, with enjoyment of the flattery, and exasperation at the inconvenience, of being turned overnight into a public figure? I must therefore try to express myself in an indirect way. I take the award of the Nobel Prize in Literature, when it is given to a poet, to be primarily an assertion of the value of poetry. I stand before you, not on my own merits, but as a symbol, for a time, of the significance of poetry.

Einstein actually *did* evade receiving his Nobel Prize. The award came long after his genius was generally acknowledged and was given not for his work on relativity but a more obscure finding—his proposal that light was sometimes a particle as well as a wave, known as the *photoelectric effect*. He claimed a prior engagement in Japan on the night of the Nobel ceremony, sent apologies to the Awards Committee, and gave an "acceptance speech" the following year in an address to the Nordic Assembly of Naturalists in Gothenburg.

He mentioned neither the photoelectric effect nor the Nobel Prize.

2 | CHOICE OR REWARD

It is February 1976 in the harbor town of Sausalito, California. The days are cold and dry. A strange redwood hut overlooks the still, gray bay. Crudely carved animals decorate its door. A beaver squeezes an accordion. An owl blows a saxophone. A dog picks at the strings of a guitar. There are no windows. Inside the hut, rock band Fleetwood Mac is recording an album called *Yesterday's Gone*. Their mood is as bleak as the weather, the atmosphere as odd as the door. The musicians hate this weird, dark studio with its strange animals. They have fired their producer. Singer Christine McVie and bass player John McVie, the Mac in the band's name, are heading for divorce. Guitarist Lindsey Buckingham and singer Stevie Nicks are riding a bronco of an affair: on again, off again, argumentative. Drummer Mick Fleetwood finds his wife in bed with his best friend. Each day at dusk they haul their emotions past the trippy critters, feast on palliative cocaine, and work past midnight. Christine McVie calls it "a cocktail party."

Fleetwood Mac survives Sausalito for a few months, then decamps to Los Angeles. Singers McVie and Nicks stay away. The Sausalito tapes are a mess. The band cancels its sold-out U.S. tour, and its record label, Warner Bros., postpones the release of *Yesterday's Gone*.

In Hollywood, forensic engineers slowly apply technical salve to the tapes and rescue the project. The band reassembles to listen and is surprised. The album is good—very good. Memories of the bickering in Sausalito inspire John McVie to change its name. He calls it *Rumours*.

Rumours is released to critical ecstasy in February 1977. It spends thirty-one weeks at the top of the *Billboard* chart, sells tens of millions of copies, wins the 1978 Grammy for album of the year, and becomes one of the bestselling records in American history, bigger than anything by the Beatles.

How to follow *Rumours*? Fleetwood Mac rents a studio in West Los Angeles, spends a million dollars, and leaves with a double album called *Tusk,* the most expensive record ever made. It gets tepid reviews; stalls at No. 4 on the charts, sells a few million copies, and sinks. Warner Bros. compares it to the rocket that was *Rumours* and declares it a failure.

A few years later, the pop band Dexys Midnight Runners met a similar fate. Dexys' big success was *Too-Rye-Ay,* an album propelled by a song called "Come On Eileen," which was the biggest-selling single of 1982 in both the United States and the United Kingdom. Like Fleetwood Mac, Dexys recorded *Too-Rye-Ay* during a storm of personal crisis: singer and bandleader Kevin Rowland and violinist Helen O'Hara were falling out of love. The album's success led to an intense world promotional tour. The musicians washed up on England's shores exhausted. Three members quit. The ones that remained went to the studio to record their next album, *Don't Stand Me Down.* It cost more money and took more time than *Too-Rye-Ay.* The photograph on the sleeve showed what was left of the band, known for wearing denim overalls, scrubbed and suited as if for job interviews. Inside were only seven songs, one of which was twelve minutes long and began with two minutes of conversation about nothing. *Don't Stand Me Down* confused reviewers, launched with no single, and did not sell. Dexys Midnight Runners would not record another album for twenty-seven years. Music business veterans call this *second album syndrome*—the one after the breakthrough that costs more, takes longer, tries harder, and fails.

Neither Fleetwood Mac nor Dexys Midnight Runners were made less creative by the emotional pressures they suffered when they recorded *Rumours* and *Too-Rye-Ay.* Like many before them, they made art from angst. But the bloom of success hides thorns of expectation. Big profits have a big price: the implied promise of more, made to a waiting, watching, wanting world.

All creators face this risk. Work we want to do is better than work we must do. Dostoyevsky bewailed the external pressure of a publisher's expectations:

This is my story: *I worked and I was tortured.* You know what it means to compose? No, thank God, you do not! I believe you have never written to order, by the yard, and have never experienced that hellish torture. Having received in advance from the *Russky Viestnik* so much money (Horror! 4,500 rubles). I fully hoped in the beginning of the year that poesy would not desert me, that the poetical idea would flash out and develop artistically towards the end of the year and that I should succeed in satisfying everyone. All through the summer and all through the autumn I selected various ideas (some of them most ingenious), but my experience enabled me always to feel beforehand the falsity, difficulty, or ephemerality of this or that idea. At last I fixed on one and began working, I wrote a great deal; but on the 4th of December I threw it all to the devil. I assure you that the novel might have been tolerable; but I got incredibly sick of it just because it was tolerable, and not *positively good*—I did not want that.

Dostoyevsky's experience was typical. Working "to order, by the yard," is less creative than working by choice.

Harry Harlow was a protégé of Lewis Terman, the father figure of the Termites we met in chapter 1. Terman's influence on Harlow was so great that he persuaded him to change his last name from "Israel" because it sounded "too Jewish." After getting a doctorate in psychology under Terman at Stanford, the newly named Harlow became a professor at the University of Wisconsin–Madison, where he renovated a vacant building and created one of the world's first primate laboratories. Some of his experiments tested the effect of reward on motivation. Harlow left puzzles consisting of a hinge held in place by bolts, pins, and bars in the monkeys' cages. The monkeys could unlock the hinge by releasing the restraints in the right order. When the monkeys opened the puzzles, Harlow reset them. After a week the monkeys had learned to open the puzzles quickly, with few mistakes. During the last five days of the experiment, one monkey opened the puzzle in less than five minutes 157 times. There was no reward: the monkeys opened the puzzles for amusement.

When Harlow introduced a reward—food—into the process, the monkeys' puzzle solving got worse. In his own words: it "tended to disrupt, not facilitate the performances of the experimental subjects." This was a surprising finding. It was one of the first times anybody noticed that external rewards could demotivate rather than invigorate.

But these were monkeys. What about people?

Theresa Amabile asked professional artists to select twenty pieces of their work, ten of which had been commissioned and ten of which had been created without a commission. A panel of independent judges assessed the merits of each piece. They consistently rated the commissioned art less creative than the self-motivated work.

In 1961, Princeton's Sam Glucksberg investigated the question of motivation using the Candle Problem. He told some people that they would win between $5 and $20, depending on how quickly they got the candle on the wall—the equivalent of between $40 and $160 in 2014 dollars. He offered other people no reward. As with Harlow's monkeys and Amabile's artists, reward had a detrimental effect on performance. People offered no reward solved the Candle Problem faster than people with a chance to win $150. Follow-up experiments by Glucksberg and others replicated these results.

The relationship between reward and motivation is not as simple as "rewards reduce performance." There are more than a hundred studies besides Amabile's and Glucksberg's. They reach no consensus. Some find that rewards help, some find that they hurt; some find that they make no difference.

Ken McGraw at the University of Mississippi offered one of the most promising hypotheses to sort out some of the mess: he wondered if tasks involving discovery were disrupted by rewards, but tasks that had one right answer, like math problems, were improved by them. In 1978, he gave students a test with ten questions. The first nine required mathematical thinking, and the tenth needed creative discovery. He offered half of the students $1.50 ($12 in 2014 dollars) if they got the problems right. He offered nothing to the other half. McGraw's results partially confirmed his idea. Reward had *no* effect on the math questions: both groups performed equally well. But it made a big difference on the creative discovery question. The subjects working for rewards took much longer to find the answer. Rewards are only a problem when open-minded thinking is required. They have a positive or neutral effect on other kinds of problem solving, but whether explicit, like Dostoyevsky's advance from the *Russky Viestnik,* or implicit, like the expectations Fleetwood Mac faced after *Rumors,* rewards clog the clockwork of creation.

Amabile explored and extended this finding with two more experiments. In the first, she asked schoolchildren to tell a story based on pictures in a book. Half the children agreed to tell a story in return for a reward—the chance to play with a Polaroid camera—and half did not. She eliminated the possibility that anticipating the reward was interfering with the children's thinking by letting them play with the camera *before* they told their story. Children in the "no reward" group got to play with the camera too, but no connection was made to the task. The children's stories were tape-recorded and judged by an independent group of teachers. The results were clear and as expected: children who'd been expecting no reward told more creative stories.

In the second experiment, Amabile introduced a new variable: *choice.* She told sixty undergraduates that they were participating in a personality test for course credit. In each case, the researcher pretended that her video recorder had broken and the experiment could not be completed. She then told members of one group, called *no choice–no reward,* that they had to make a collage instead. She told subjects in another group, called *no choice–reward,* that they had to make a collage but would be paid $2. She asked people in a third group, *choice–no reward,* if they would mind making a collage but did not offer payment. She asked members of the fourth group, *choice–reward,* if they would mind making a collage for $2. For added emphasis, the reward groups worked with two dollar bills in front of them. An independent panel of experts judged the collages. In this experiment, reward *did* lead to the most creative work—by the *choice–reward* group. But the *least* creative work was also caused by reward—it came from the *no choice–reward* group. Both *no reward* groups scored in the middle, regardless of whether or not they had been given a choice. In creative work, choice transforms the role of reward. The least creative group's problem was easily diagnosed: members of the *no choice–reward* group reported feeling the most pressure.

No *choice–reward* is the condition most of us are in when we go to work.

3 | THE CROSSROADS

People in America's Deep South tell a story about a musician named Robert Johnson. They say one night when the crickets were quiet and clouds curtained the moon, Johnson stole out of his bed at Will Dockery's plantation cradling his guitar. He followed the Sunflower River by the light of the stars until it brought him to a crossroads in the dust bowl where a tall, dark figure stood waiting. The figure took Johnson's guitar with hands strange and large, tuned it, then played it so the strings wailed and wept with mortal emotion, making a music no man had heard before. When he finished playing, the stranger revealed his identity: he was the Devil. The Devil offered Johnson a trade: the sound of the guitar for Johnson's soul. Johnson took the deal and became the greatest guitarist who ever lived, playing the Devil's music, which was called "the blues," all along the Mississippi Delta until he became legend. After six years, the Devil called in his due and took Robert Johnson's soul. Johnson was twenty-seven.

The story is neither completely true nor completely false. There *was* a man named Robert Johnson. He *did* play the blues along the

Mississippi Delta for six years. He was one of the greatest guitarists who ever lived. His legacy includes blues, rock, and metal. He died at twenty-seven. He did not make a deal with the Devil, but he did come to a crossroads where he had to make a deal with himself. Johnson married at nineteen and, despite his talent as a musician, planned a stable life as a farmer and father. It was only when his wife, Virginia, died in childbirth that he resolved to do what others described as "sell his soul to the Devil" and fully commit to playing the blues.

The story that arose around Robert Johnson's life and talent is partly due to his early death; partly due to his song "Cross Road Blues," which tells a tale about failing to hitch a ride, not a deal with the Devil; and mostly a mixture of ancient German legend and African American myth.

The German legend is the story of Faust, which dates back at least as far as the sixteenth century. It comes in many flavors but has one common theme. Faust is a learned man, typically a doctor, who yearns for knowledge and magical power. He calls upon the Devil and strikes a bargain. Faust gets knowledge and magic, and the Devil gets Faust's soul. Faust enjoys his powers until the Devil returns and takes him to Hell.

According to hoodoo, the folk mythology of African slaves, you can acquire special skills if you meet a dark stranger at a crossroads in the dead of night. The voodoo traditions of Haiti and Louisiana also reserve a special role for the crossroads: they connect the spiritual world to the material world and are guarded by a gatekeeper called Papa Legba. Unlike the legend of Faust, this stranger at the crossroads demands no price.

Robert Johnson's story blends these two mythic archetypes to illuminate a deeper truth: there comes a point in every creative life, no matter what the discipline, when success depends upon committing completely. The commitment has a high price: we must devote ourselves almost entirely to our creative goal. We must say no to distraction when we want to say yes. We must work when we do not know what to do. We must return to our creation every day without excuse. We must continue when we fail.

If any devil is involved, he is not the one demanding commitment. Whatever your higher power, whether God, Allah, Jehovah, Buddha, or the greater good of humanity, this is whom you serve when you commit to a life of creation. What is diabolical is squandering your talents. We sell our soul when we waste our time. We drive neither ourselves nor our world forward if we choose idling over inventing.

When Robert Johnson came to the crossroads at midnight, it was temptation that said, Do not practice, do not play, do not write, do not stretch your hands across the frets until they ache, do not press your fingers into the strings until they bleed, do not play to empty chairs and chattering drunks who boo, do not perfect your music, do not train your voice, do not lie awake with your lyrics until every word sounds right, do not study the skill of every great player you hear, do not invest your every breathing, waking minute pursuing your God-given mission to create. Take it easy, mourn your wife and child, get some rest, have a drink, play some cards, hang with your friends — they do not spend all day and night messing with guitars and music.

And Robert Johnson looked at temptation and said *no*. Then he took his guitar to the Mississippi Delta and for six years played music so great it changed the world, music so great it inspired every guitarist that followed, music so great we are discussing him now not because our topic is guitars or even music but because his story breathes life into the true meaning of creative commitment.

If you are fully immersed in your creative life and the crossroads has long left your rearview mirror, be affirmed. The friends, mothers, fathers, therapists, colleagues, ex-boyfriends, ex-girlfriends, ex-husbands, and ex-wives who said you were crazy and you work too hard and you will never make it and you need more balance were wrong, as are the ones who still do.

If you have not yet reached the crossroads, look around. It is here now. That stranger over there is waiting for the chance to offer you an endless supply of reasons why you should not create a thing.

All he wants in return is your soul.

4 | TWO TRUTHS OF HARRY BLOCK

Some say there is a condition called "writer's block"—a paralysis that prevents people from creating. Writer's block is alleged to cause depression and anxiety. Some researchers have speculated that it has neurological causes. One has even attributed it to "cramping" in the brain. But no one has found any evidence that writer's block is real. It is the inevitable underbelly of that other unproven phenomenon, the aha! moment. If you can create only when you are inspired, then you cannot create when you are not inspired; therefore, creating can be blocked.

Woody Allen makes fun of writer's block. He wrote a play called *Writer's Block,* and he wrote, directed, and starred in a movie, *Deconstructing Harry,* in which the protagonist, Harry Block, tells his therapist: "For the first time in my life I experience writer's block. . . . Now this, to me, is unheard of. . . . I start these short stories and I can't finish them. . . . I can't get into my novel at all . . . because I took an advance."

Allen took the role of Harry, but only as a last resort, two weeks before filming began, because other actors, including Robert De Niro, Dustin Hoffman, Elliott Gould, Albert Brooks, and Dennis Hopper were not available. Allen was afraid people would assume Harry Block was autobiographical, when, in fact, he is antithetical: "He's a New York Jewish writer—that's me—but he's a writer with writer's block—that immediately disqualifies me."

Writer's block immediately disqualifies Harry Block from being Woody Allen because Woody Allen is one of the most productive film makers of his—and possibly any—generation. Between 1965 and 2014, Allen was credited in more than sixty-six films as a director, writer, or actor—often all three. To take writing alone: Allen has written forty-nine full-length theatrical films, eight stage plays, two television films, and two short films in less than sixty years, a rate of over a script a year, despite directing and acting in movies at about the same rate. The only other moviemakers who come close are Ingmar Bergman, who wrote or directed fifty-five films in fifty-nine years but did not act in any of them, and directors from the "factory" studio system of the 1930s, like John Ford, who directed 140 films, sixty-two of them silent, in fifty-one years but did not write or act in any.

Allen's productivity tells two truths about writer's block. The first is about the importance of time:

> I never like to let any time go unused. When I walk somewhere in the morning, I still plan what I'm going to think about, which problem I'm going to tackle. I may say, this morning I'm going to think of titles. When I get in the shower in the morning, I try to use that time. So much of my time is spent thinking because that's the only way to attack these writing problems.

A victim of "writer's block" is *not* unable to write. He or she can still hold the pen, can still press the keys on the typewriter, can still power up the word processor. The only thing a writer suffering from writer's block cannot do is write something they think is good. The condition is not writer's block, it is write-something-I-think-is-good block. The cure is self-evident: write something you think is bad. Writer's block is the mistake of believing in constant peak performance. Peaks cannot be constant; they are, by definition, exceptional. You will have good days and less good days, but the only bad work you can do is the work you do not do. Great creators work whether they feel like it or not, whether they are in the mood or not, whether they are inspired or not. Be chronic, not acute. Success doesn't strike; it accumulates.

Woody Allen learned this early, writing jokes for television, saying: "You couldn't sit in a room and wait for your muse to come and tickle you. Monday morning came, there was a dress rehearsal Thursday, you had to get that thing written. And it was grueling, but you learned to write." And:

> Writing doesn't come easy, it's agonizing work, very hard, and you have to break your neck doing it. I read many years later that Tolstoy said, in effect, "You have to dip your pen in blood." I used to get at it early in the morning and work at it and stay at it and write and rewrite and rethink and tear up my stuff and start over again. I came up with such a hard-line approach—I never waited for inspiration; I always had to go in and do it. You know, you gotta force it.

Writer's block is not the same as getting stuck. Everybody gets stuck. The myth of writer's block may exist partly because not everybody knows how to get unstuck. Allen:

> I've found over the years that any momentary change stimulates a fresh burst of mental energy. So if I'm in this room and then I go into the other room, it helps me. If I go outside to the street, it's a huge help. If I go up and take a shower it's a big help. So I sometimes take extra showers. I'll be in the living room and at an impasse and what will help me is to go upstairs and take a shower. It breaks up everything and relaxes me. I go out on my terrace a lot. One of the best things about my apartment is that it's got a long terrace and I've paced it a million times writing movies. It's such a help to change the atmosphere.

Allen's second truth about writer's block is a confirmation that intrinsic motivation is the only motivation. Inspirational lightning bolts are external—they come from without and are beyond our control. The power to create must come from within. Writer's block is waiting—waiting for something outside of yourself—and just a shinier way to say "procrastination."

Much of the paralysis of writer's block comes from worrying what others will think: write-something-I-think-is-good block is often rooted in write-something-somebody-else-will-think-is-good block. Woody Allen's indifference to other people's opinion about his work is one big reason why he is so productive. He is even indifferent to what other people think of his productivity: "Longevity is an achievement, yes, but the achievement that I'm going for is to try to make a great film. That has eluded me over the decades."

Not only does Allen not go to awards shows, he does not read any of his reviews and has not even been to see all his own movies. The work, specifically the satisfaction he takes from it for himself, is its own statuette: "When you actually sit down to write, it's like eating the meal you've spent all day in the kitchen cooking."

Cook to eat, not to serve.

5 | THE OTHER HALF OF KNOWING

The largest island in the Philippine archipelago is Luzon, which reaches like a wing from Manila toward China and Taiwan. In the east, the Mingan Mountains climb to wild green peaks six thousand feet high. Until the eighteenth century, these mountains kept a secret: an indigenous people called the Abilaos or Italons or, most commonly in English, the Ilongots.

As recently as fifty years ago, the Ilongots had a ferocious reputation. *Popular Science* described them as "savages, treacherous murderers, and wholly untamable." They were known to be headhunters—people who murdered and decapitated their neighbors, keeping their victims' heads, and sometimes heart and lungs, as trophies.

In 1967, Michelle Rosaldo, an anthropologist from New York, went to live with the Ilongots. The Ilongots were doing much less head-hunting in the 1960s, but this was still a brave step. The last anthro-pologist to live with the Ilongots, William Jones, had been there less than a year when three Ilongots, including the man he shared a hut with, killed him with knives and spears.

What Rosaldo found was a culture with a distinct view of human nature. The Ilongots believe that everything human is the result of two psychological forces: *bēya*, or knowledge, and *liget*, or passion. Success in life comes from tempering passion with knowledge. Passion with knowledge brings creation and love; passion without knowledge brings destruction and hate. Passion, they believe, is innate and dwells in the heart. Knowledge is instilled and found in the head. The purpose of each Ilongot's life was to develop the knowledge they needed to focus their passion into creation for the common good. Headhunting and other forms of violence were the result of too much passion and not enough knowledge. Amazed, Rosaldo captured the Ilongots' insights in a book, *Knowledge and Passion,* now an iconic work of anthropology.

Stories like Woody Allen's and experiments like Teresa Amabile's show us that passion matters but not what passion *is*. The wisdom of the Ilongots fills that gap. Passion is the most extreme state of choice without reward. Or, rather, it is *its own reward,* an energy that is indif-ferent to outcomes, even when they include missed sleep, becoming poor, losing your friends, bleeding and bruising, even death.

This is not a new definition. The word "passion" comes from the Latin *passio,* for "suffering." In 1677, Dutch philosopher Baruch Spi-noza defined passion as a negative state in his masterwork, *Ethica*

Ordine Geometrico Demonstrata, or *The Ethics:* "The force of any passion or emotion can overcome the rest of a man's activities or power, so that the emotion becomes obstinately fixed to him."

Spinoza thought passion was the opposite of reason—a force for madness. French philosopher René Descartes had a different view: "We can't be misled by passions, because they are so close, so internal to our soul, that it can't possibly feel them unless they are truly as it feels them to be. Even when asleep and dreaming we can't feel sad or moved by any other passion unless the soul truly has this passion within it."

Or: passion is the voice of the soul.

The two definitions of passion dueled until the twentieth century, when the positive view became more popular. But is passion always good? The Ilongots show us the answer. Passion is energy; if it does not create, it harms.

6 | ADDICTION, SORT OF

As the Ilongots and their headless victims know, passion that does not create destroys. We are all creative, and whether we have found it yet or not, we all have passion. But so many of us, for one reason or another, do not put our passion into action. Unfulfilled passion creates a cavity between our present and our potential—a void that can drip with destruction and despair. It stagnates. It manifests as might-have-beens. If we do not chase our dreams, they will pursue us as nightmares. Unfulfilled passion creates addicts and criminals.

Daquan Lawrence celebrated his sixteenth birthday incarcerated at the Elliot Hillside Detention Center in Roxbury, part of Boston, Massachusetts. His parents were drug addicts. His Nana Charlesetta rescued him from their home when he was five. He was arrested for the first time at thirteen, for dealing marijuana and crack on the streets of Mattapan, Roxbury's troubled neighbor, known in Boston as "Murderpan." (Roxbury itself has a marginally better reputation. Bostonians only call *that* part of town "Roachbury.") Lawrence careened from one prison to the next for the rest of his childhood, known to all, including himself, as a repeat offender and maker of trouble.

Then, soon after that jailhouse sweet sixteen, a pipe-thin stranger came to Elliot Hillside. His name was Oliver Jacobson. He carried heavy black boxes into the detention center's staff room. Lawrence peered shyly through the door. He saw Jacobson unpacking a piano. Trails of cables connected microphones, keyboards, speakers, and headphones.

Encouraged by Jacobson, Lawrence approached a microphone and tried to rap. It was a moment of revelation for all who saw it: Lawrence, the hopeless young pusher on a treadmill of crime and punishment or worse, had the gift the hip-hop world calls "flow." His rap was fluid, on time, and in tune. He ad-libbed—or "freestyled"—using a range of poetic tricks, from rhyme and repetition to assonance and alliteration:

> *It's the strive from inside that reveals the pride,*
> *But the message from the sky that shows me the guide,*
> *We are leaders, overachievers,*
> *Stuck once in the vision and precision of believers,*
> *Keep looking up to the sky, you keep flowin',*
> *Never stop in the dark, you are glowing.*

Lawrence spent months writing songs with Jacobson. He gave himself the rap name "True." He studied acting, playing Romeo in *Romeo and Juliet* and Othello in *Othello*.

When he got out of jail, at seventeen, he took his first-ever job to pay for acting school—as a door-to-door salesman for an energy company. He passed his General Educational Development test, earning the equivalent of a high school diploma. He started thinking about college and told the *Boston Globe,* "The arts taught me to have a direction, to have a goal, be the best you can be. It feels like I've been productive in every way. I feel like right now, it's real right for me. It's meaningful."

Daquan's story is not unusual. Rappers have a reputation for becoming criminals, but criminals more frequently become rappers, or musicians in other genres, or writers, actors, artists, or creators of some other kind. In 1985, a seventeen-year-old crack cocaine dealer named Shawn Corey Carter borrowed a gun and shot his older brother in an argument about jewelry; in 1999 he was arrested and tried for allegedly stabbing a man in the stomach in a New York nightclub. Carter pleaded guilty to a misdemeanor and was given three years' probation. It was a turning point. He said, "I vowed to never allow myself to be in a situation like that again." Today, Carter is better known as a rapper named Jay-Z. By 2013, after twenty years of success in music and business, he had a personal fortune of around half a billion dollars.

Music has diverted children from crime all over the world. Israel has a program called Music Is the Answer; Australia's Children's Music Foundation has a Disadvantaged Teens program; Oliver Jacobson, Daquan Lawrence's music teacher, was a volunteer for the U.S. nonprofit Genuine Voices; and in Britain, the Irene Taylor Trust, a charity, operates a program called Music in Prisons. In an evaluation of one of its programs, the Irene Taylor Trust claimed that prisoners were 94 percent less likely to commit a crime during the program and 58 percent less likely to commit a crime in the six months after completing it. These numbers are too good to be true — the data is sparse and the research poorly controlled. It would be wrong to say that a few months of music school ends a life of crime: Daquan Lawrence continued to deal drugs, and get caught doing it, for several years after he started to rap. But all the good outcomes make the truth obvious: the more we create, the less we destroy.

We are inclined to regard passion as positive and addiction as negative, but they are indistinguishable apart from their outcomes. Addiction destroys, passion creates, and that is the only difference between them. In the 1950s, George "Shotgun" Shuba hit baseballs for the Brooklyn Dodgers. One night after he retired, Shuba sat in his basement with sportswriter Roger Kahn, drinking cognac and talking about the game. Shuba described how as a boy he had practiced by hanging a length of knotted rope in his backyard and hitting it with a weighted bat. Then, old and slightly drunk, he demonstrated. From a case on the wall he took a bat weighted with lead and prepared to hit an old knotted rope as if it were a ball. Kahn described what happened next:

> The swing was beautiful, and grunting softly he whipped the bat into the clumped string. Level and swift, the bat parted the air and made a whining sound. Again Shuba swung and again, controlled and terribly hard. It was the hardest swing I ever saw that close.
>
> I said, "You're a natural."
>
> "Ah," Shuba said. "You talk like a sportswriter."
>
> He went to the file and pulled out a chart, marked with Xs.
>
> "In the winters," he said, "for fifteen years after loading potatoes or anything else, even when I was in the majors, I'd swing six hundred times. Every night, and after sixty I'd make an X. Ten Xs and I had my six hundred swings. Then I could go to bed.
>
> "You call that natural? I swung a 44-ounce bat 600 times a night, 4,200 times a week, 46,200 swings every winter."

The secret of Shuba's swing was what psychologist William Glasser later called "positive addiction." Shuba was so passionate about baseball that he acted like an addict. He could not sleep unless he had swung his bat six hundred times. His addiction, or passion, became his career.

One way or another, your passion will out. Use it as the courage to create.

Passion must be structured by process. Woody Allen starts with a drawer full of pieces of paper, many torn from matchbooks and magazine corners, all little patches of possibility:

> I'll start with scraps and things that are written on hotel things, and I'll, you know, ponder these things, I'll pull these out and I'll dump them here like this on the bed. I have got to go through this all the time, and every time I start a project I sit here like this, and I look. A note here is "A man inherits all the magic tricks of a great magician." Now that's all I have there, but I could see a story, forming where some little jerk like myself at an auction or at some opportunity buys all those illusions and you know boxes and guillotines and things and it leading me to some kind of interesting adventure going into one of those boxes and maybe suddenly showing up in a different time frame or a different country or in a different place altogether. I'll spend an hour thinking of that and it'll go no place and I'll go on to the next one.

The three most destructive words in the English language may be *before I begin*.

Oscar-winning screenwriter Charlie Kaufman: "To begin, to begin. How to start? I'm hungry. I should get coffee. Coffee would help me think. I should write something first, then I'll reward myself with coffee. Coffee and a muffin. OK, so I need to establish the themes. Maybe banana-nut. That's a good muffin."

The only thing we do before we begin is fail to begin. Whatever form our failure takes, be it a banana-nut muffin, a tidier sock drawer, or a bag of new stationery, it is the same thing: a non-beginning, complete with that dead car sound, all click, no ignition. Having resisted the temptation of others, we must also resist the temptation of us.

The best way to begin is the same as the best way to swim in the sea. No tiptoes. No wading. Go under. Get wet and cold from scalp to sole. Splutter up salt, push the hair from your brow, then stroke and stroke again. Feel the chill change. Do not look back or think ahead. Just go.

In the beginning, all that matters is how much clay you throw on the wheel. Go for as many hours as you can. Repeat every day possible until you die.

The first beginning will feel wrong. We are not used to being with ourselves uninterrupted. We do not know the way first things look. We have imagined our creations finished but never begun. A thing begun is less right than wrong, more flaw than finesse, all problem and no solution. Nothing begins good, but everything good begins. Everything can be revised, erased, or rearranged later. The courage of creation is making bad beginnings.

Russian composer Igor Stravinsky, one of the great innovators of twentieth-century music, played a Bach fugue on the piano every morning. He started every day like this for years. Then he worked for ten hours. Before lunch he composed. After lunch he orchestrated and transcribed. He did not wait for inspiration. He said, "Work brings inspiration if inspiration is not discernible in the beginning."

Ritual is optional, but consistency is not. Creating requires regular hours of solitude. Time is your main ingredient, so use the highest-quality time to create.

At first, creating for an hour is hard. Every five minutes our mind itches for interruption: to stretch, get coffee, check e-mail, pet the dog. We indulge an urge for research, and before we know it we have Googled three links away from where we started and are reminding ourselves of the name of Bill Cosby's wife in *The Cosby Show* (it was Clair) or learning what sound a giraffe makes (giraffes are generally quiet, but they sometimes cough, bellow, snort, bleat, moo, and mew). This is the candy we give ourselves.

What solitude creates interruption destroys. Science describes the destruction unequivocally. Many experiments show the same things: interruption slows us down. No matter how little time is stolen by interruption, we lose even more time reconnecting to our work. Interruption causes twice as many mistakes. Interruption makes us angry. Interruption makes us anxious. These effects are the same among men and women. Creation knows no multitasking.

Interruption, unfortunately, is also addictive. We live in an interruption culture, and it conditions us to crave interruption. Say no to the itch. More "no's" equals fewer itches. The mind is a muscle that starts soft but becomes long and lean with use. The more we focus, the stronger it gets. After that first difficult hour, several seem easy. Then we not only work for hours but also feel wrong if we do not. A change comes. The itch is not for interruption but for concentration.

As we sit with our pen poised above the page we aim to turn into a novel, scientific paper, work of art, patent, poem, or business plan, we can feel paralyzed—and that's only if we can summon up the courage to sit in the first place. While knowing that this is a natural and normal part of the creative process may ease our minds a little, it may not make us more productive. We look around for inspiration. This is the right thing to do, only much of art lies not in what we see but in what we don't. When we envy the perfect creations of others, what we do not see, what we by definition *cannot* see, and what we may also forget when we look back at successful creations of our own, is everything that got thrown away, that failed, that didn't make the cut. When we look at a perfect page, we should put it not on a pedestal but on a pile of imperfect pages, all balled or torn, some of them truly awful, created only to be thrown away. This trash is not failure but foundation, and the perfect page is its progeny.

The most creative force we can conceive of is not us, it is what created us, and we can learn from it. Call it God or evolution; it is undeniably a brutal editor. It destroys almost all of what it makes, through death, extinction, or simple failure to reproduce or be produced, and selects only the best of what is left for survival. Creation is selection.

Everything, whether nature or culture, was created by this process. Every peach, every orchid, every starling, like every successful act of art, or science, or engineering, or business, is made of a thousand failures and extinctions. Creation is selection, iteration, and rejection.

Good writing is bad writing well edited; a good hypothesis is whatever is left after many experiments fail; good cooking is the result of choosing, chopping, skinning, shelling, and reducing; a great movie has as much to do with what ends up on the cutting room floor as what does not. To succeed in the art of new, we must fail freely and frequently. The empty canvas must not stay empty. We have to plunge into it.

What we produce when we do will be bad, or at least not as good as it will become. This is natural. We must learn to be at ease with it. Whenever we begin to invent or create or conceive, whenever we begin to make something new, our heads fill with advocates of same, holding censor's pencils, babbling criticism. We recognize most of them. They are the ghosts of hecklers, judgers, investors, and reviewers past, present, and future, personifications created by our evolved instinct for keeping things the way they are, manifesting to stay our hand and save us from the perils of new.

These characters—all us in disguise, of course—should be welcomed, not rejected. They are important and useful, but they have arrived too early. The time for critical assessment—their time—comes later. For now they must be shown to a room in our mind where they can wait, unheard, until needed for editing, assessing, and redrafting. Otherwise they will not only paralyze us, they will drain our imagination. It takes a lot of energy to script and voice all those naysayers— energy we need for the task at hand.

The same is true of their opposites. Sometimes inner critics are replaced by cheerleaders of new who urge us on with fantasies of fame and glamour. They imagine the first bad stanza we write bringing down the house on Broadway. They script our Nobel acceptance speech while we are drafting the title of our scientific paper. They rehearse the anecdotes we will share on the couches of chat shows while we are writing the first page of our novel. For these voices, anything new we make, or even conceive of, is perfect. Show them to the waiting room, too.

Almost nothing we create will be good the first time. It will seldom be bad. It will probably be a dull shade of average. The main virtue of a first sketch is that it breaks the blank page. It is a spark of life in the swamp, beautiful if only because it is a beginning.

And, somehow, long after the beginning and far into an endless middle, something takes shape. After the tenth prototype, the hundredth experiment, or the thousandth page, there is enough material to enable selection. All that clay thrown on the wheel has the potential to be more than new. It has the potential to be good.

This is the time for those advocates of same, our inner critics and judges, to be let through the door they have been pounding on for so long. They have been eavesdropping all the while and are ready to attack the work with blue pencils as sharp as teeth and claws. Let them be loud. Let them brutally scrutinize the data, or the draft, or the sketch and cut out anything and everything that doesn't need to be there. Selection is a bloody process. Beautiful work, maybe months in the making, is culled in moments.

This is the hardest part of all. We are the sum of our time and dreams and deeds, and our art is all three. Abandoning an idea can seem like losing a limb. But it is not nearly as serious, and it has to be done. The herd must be thinned or it faces extinction, and any new work that does not suffer selection faces an equivalent fate: it is unlikely to pass peer review, or be produced or patented, exhibited, or published. The world will always be more hostile to our work than we are. Ruthless selection gives it less to work with.

When the frenzy is over and only our fittest work, our very best new, has survived, it is time to begin again. The agents of same, sated for now, must retreat so that whatever is left, however slight, can reproduce and grow into a second draft, another prototype, a changed experiment, a rewritten song, stronger and better adapted.

And so it goes on. No eurekas or flashes of inspiration. Innovation is whatever remains when all our failures are removed. The only way to work is to accept our urge to create *and* our desire to keep things the same and make both pull in our favor. The art of new, and perhaps the art of happiness, is not absolute victory for either new or old but balance between them. Birds do not defy gravity or let it bind them to the ground. They use it to fly.

8 | FROM E TO F

Why do more when you can do less? Woody Allen has pondered that, too: "Why opt for a life of grueling work? You delude yourself that there's a reason to lead a productive life of work and struggle and perfection of one's profession or art. My ambitions or my pretensions—to which I freely admit—are not to gain power. I only want to make something that will entertain people, and I'm stretching myself to do it."

New is difference, so difference makes new. When we create, we harvest what is uniquely ours, our speck of special, our very selves, shaped by our genes, by the life that courses over us daily, and, for those of us who have them, our God or gods. We each bring difference to the world. It is inside us from birth to death. Every parent knows that their child is like no other, made from a recipe of talents, tendencies, tics, and loves all his or her own. My first child loved snow before she could walk. My second rejected his first snowfall by crying for a chai latte. He was not yet two. What makes us prefer a chai latte to snow before we are two? Something innate. No matter how many billions breathe

the air of this earth, *you* bear something that has never before been borne and will not be borne again: a gift to be given not kept.

We may not write symphonies or discover laws of science, but new is in all of us. There is a bakery in my old neighborhood in Los Angeles. It is tiny, forty seats or fewer. A woman called Annie Miler created it in 2000. Annie is a pastry chef. She makes blueberry muffins, butterscotch brownies, and grilled cheese sandwiches. The bakery's interior is artful, tasteful, and personal. You can see Annie growing up in pictures hanging along the wall. In the first she is a little redheaded girl shyly displaying an early cake; by the last she stands with her team on her bakery's opening day. Annie's baking binds her community. Her store is where neighbors meet to pet each other's dogs and share small talk over the tang of espresso. The seasons change with the fruit in her tarts and the flavors of her soups. People go to her bakery to kick-start their day, to have first dates, and to salve the pains of life.

Annie's place probably sounds like a place you know. There are many people who, like Annie, have built boutiques, cafés, florists, delis, and thousands of other community businesses that go beyond mere franchises or cookie-cutter stores and have new and unique details because they are reflections of what is new and unique about their creators.

Be like Woody Allen and Annie Miler. Make passion the gas in your tank.

CHAPTER 8

CREATING ORGANIZATIONS

1 | KELLY

In January 1944, Milo Burcham strolled across an airstrip in California's Mojave Desert and climbed into a plane called *Lulu Belle*. *Lulu Belle* looked like an insect: shiny green with stubby wings and no propellers. A crowd of men, swaddled in overcoats, watched in silence. Burcham started the engine, a de Havilland Goblin from England— the only one of its kind in the world—shot a brief glance of mischief at his audience, and then accelerated into the sky. When he reached 502 miles an hour, he dropped *Lulu Belle* low and flew her so close to the men that he could see them startle. They were still watching in silence when he landed and opened the cockpit. He pushed himself out, wearing his best just-another-walk-in-the-park face, fighting a grin and winning, until the men whooped and ran toward him, hollering and clapping as if they had never seen a plane before. Burcham cracked a smile as wide as the sky. It was *Lulu Belle*'s first flight. No American plane had ever flown so fast.

Lulu Belle's official name was the "Lockheed P-80 Shooting Star."

She was the first fighter jet in the United States military. It was forty years since the Wright brothers' first flight at Kitty Hawk and 143 days since the P-80 had been conceived.

If creating is best done alone by people with intrinsic motivation and free choice, how do creative *teams* work? How can anyone build an *organization* that creates?

The team that built the P-80 at the height of World War II faced a hard problem: build a jet-powered airplane that can fight, and build it fast. The urgency was a matter of life and death. In 1943, British code breakers had discovered something horrifying: Hitler's engineers had built a jet-powered fighter plane with a top speed of 600 miles per hour. The plane, called the Messerschmitt Me 262 Schwalbe, or Swallow, was agile and highly maneuverable, despite being armed with four machine guns, rockets, and, if necessary, bombs. It was in mass production. It would be raining death on Europe by early 1944. The Nazis were winning a new kind of war—a war from the sky, using technology that had been inconceivable just a few years earlier.

The man who led the team to counter the threat of the Messerschmitt was Clarence Johnson, an engineer known to all as "Kelly." The urgency and complexity of the challenge were not Johnson's only problems—the United States government was also sure that German spies were listening to its communications. Johnson had to build a secret lab using old boxes and a tent rented from a circus and hide it next to a wind tunnel at the Lockheed plant in Burbank, California. He could not hire secretaries or janitors, and his engineers could not tell anyone, not even their families, what they were doing. One engineer called the place the "Skonk Works," after a factory that ground skunks and shoes into oil in a popular comic strip called *Li'l Abner*. The name stuck until long after the war, and when the secrecy lifted, the comic's publisher made Lockheed change the name. From then on the operation, technically Lockheed's advanced projects division, was called the "Skunk Works."

The circumstances forced on Kelly Johnson seemed adverse but turned out to be fortuitous—he discovered that a small, isolated, highly motivated group is the best kind of team for creation. The United States military gave Johnson and his team six months to design America's first jet fighter. They needed fewer than five. The P-80 was the first plane developed by the engineers at the Skunk Works, and they went on to invent the supersonic F-104 Star Fighter; the U-2 surveillance plane; the Blackbird surveillance plane, which flew at three times the speed of sound; and aircraft that could evade radar detection. In addition to creating planes, Johnson created something else: a model organization for achieving the impossible quickly.

2 | SHOW ME

Kelly Johnson started working at Lockheed in 1933. It was a small airplane manufacturer with only five engineers, restructuring after bankruptcy and struggling to compete with two much larger companies: Boeing and Douglas (later McDonnell Douglas). Johnson's first day at Lockheed could have been his last. He had been hired in part because, as a student at the University of Michigan, he had helped test Lockheed's new Model 10 Electra all-metal airplane in the university's wind tunnel. His professor, Edward Stalker, the head of Michigan's department of aeronautical engineering, had given the Electra a good report. Johnson disagreed. On day one at Lockheed, the twenty-three-year-old, who had only just received his graduate degree in aeronautics and had been hired not as an engineer but to make technical drawings, said so:

> I announced that the new airplane, the first designed by the reorganized company and the one on which its hopes for the future were based, was not a good design, actually was unstable. They were somewhat shaken. It's not the conventional way to begin

employment. It was, in fact, very presumptuous of me to criticize my professors and experienced designers.

There are few companies today where this would be a good career move. In the 1930s, there were probably even fewer. What happened next almost explains Lockheed's success all by itself.

Johnson's boss was Hall Hibbard, Lockheed's chief engineer. Hibbard's aeronautics degree was from the Massachusetts Institute of Technology, then and now one of the world's greatest engineering schools. He wanted what he called "new young blood"—people who were "fresh out of school with newer ideas." Hibbard said, "When Johnson told me that the new airplane we had just sent in to the university wind tunnel was no good, and it was unstable in all directions, I was a little bit taken aback. And I wasn't so sure that we ought to hire the guy. But then I thought better of it. After all, he came from a good school and seemed to be intelligent. So, I thought, let's take a chance."

Instead of firing Johnson for impudence, Hibbard sent him on his first business trip, saying, "Kelly, you've criticized this wind-tunnel report on the Electra signed by two very knowledgeable people. Why don't you go back and see if you can do any better with the airplane?"

Johnson drove twenty-four hundred miles to Michigan with a model of the Electra balanced in the back of his car. He tested it in the wind tunnel seventy-two times, until he solved the problem with an unusual "twin" tail that had a fin on each side of the aircraft and nothing in the center.

Hibbard's response to the new idea was to work late writing Johnson a letter:

Dear Johnson,

You will have to excuse the typing as I am writing here at the factory tonight and this typewriter certainly is not much good.

You may be sure that there was a big celebration around these parts when we got your wires telling about the new find and how simple the solution really was. It is apparently a rather important discovery and I think it is a fine thing that you should be the one to find out the secret. Needless to say, the addition of these parts is a very easy matter; and I think that we shall wait until you get back perhaps before we do much along that line.

Well, I guess I'll quit now. You will be quite surprised at the Electra when you get here, I think. It is coming along quite well.

Sincerely,
Hibbard.

When Johnson returned to Lockheed, he found that he had been promoted. He was now Lockheed's sixth engineer.

The story of the Skunk Works, America's first jet fighter, its supersonic aircraft, stealth technology, and whatever may follow that started with this one act. In almost any other company, or talking to almost any other manager, Johnson would have been laughed out of the room, and possibly out of his job. That was Hibbard's first instinct, too. But Hibbard had a rare trait: he was intellectually secure.

Intellectually secure people do not need to show anyone how smart they are. They are empirical and seek truth. Intellectually *in*secure people need to show *everyone* how smart they are. They are egotistical and seek triumph.

Intellectual security is not related to intellect. People who are more skilled with their hands than their minds are often intellectually secure. They know what they know and enjoy people who know more. Brilliant people are usually intellectually secure, too—and for the same reason.

Intellectual insecurity is most commonly found in the rest of us: people who are neither nonintellectual nor extremely intellectual. Not only are we the vast majority—we are also the people most likely to be made managers. People who are mainly skilled with their hands are no more interested in management than are Nobel laureates. As a result, most managers and executives are intellectually insecure. Hall Hibbard was unusual, and he was in the right place at the right time.

Hibbard's response to his new employee's bold claim that Lockheed's plane was a lemon was the perfect one. One of the most powerful things any manager can say is "Show me."

Frank Filipetti, a producer for musicians including Foreigner, Kiss, Barbra Streisand, George Michael, and James Taylor, uses "show me" to manage creative conflict in the recording studio:

> When you're dealing with a creative process, there's always ego involved. I have one philosophy: I never want to get into an argument about, or discuss, how something is going to sound. I've had people sit there and tell me why putting the backgrounds in the first chorus isn't going to work, and they'll expound on that for thirty minutes, when all you've got to do is play the damn thing, and then you'll hear it. And more times than not, everybody agrees, once they hear it. But they'll sit there and they'll argue this thing out without listening to it. You can intellectualize all this stuff until you're blue in the face, but the end result is the way it sounds, and that can really surprise you sometimes. There have been times when I thought I was absolutely right, and then I listen to it, and I have to admit, "That actually sounds pretty good." Once you get to that stage where you say, "Let's just play it," it's really amazing how everybody kind of hears the same thing all of a sudden. And it takes that ego thing out of it, too.

Hibbard's letter was the equivalent of "That actually sounds pretty good." It meant so much to Johnson that he kept it his entire life.

In November 1960, Robert Galambos figured something out. He said out loud, to no one in particular, "I know how the brain works."

A week later, Galambos presented his idea to David Rioch, his manager for the past ten years. The meeting went badly. Rioch did not say, "Show me." Instead, Galambos's idea made Rioch angry. He ordered Galambos not to talk about it in public, or to write about it, and predicted that his career was over. And it nearly was: within months, Galambos was looking for a new job.

Both men were neuroscientists at the Walter Reed Army Institute of Research, in Silver Spring, Maryland. They had worked together closely for a decade, trying to understand how the brain works and how to repair it. They and their colleagues had made Walter Reed one of the world's most respected and prestigious centers of neuroscience. Galambos, then forty-six years old, was more than just an accomplished neuroscientist—he was also a famous one. When he was a researcher at Harvard, he had for the first time conclusively proved, with collaborator Donald Griffin, that bats use echolocation to "see in the dark"—a radical finding that was not immediately accepted by experts but that we now take for granted. Despite this pedigree, and their long history of successfully working together, Rioch quickly forced Galambos out of his job because of his new idea. Six months later, Galambos left Walter Reed forever.

Galambos's idea was apparently simple: he hypothesized that cells called "glia" are crucial to brain function. Forty percent of all brain cells are glia, but in 1960 it was assumed that they didn't do anything but hold the other, more important cells together and perhaps support and protect them. This assumption was built right into their name: the word *glia* is medieval Greek for "glue."

Rioch's problem with Galambos's idea dates back to a nineteenth-century Spaniard, Santiago Ramón y Cajal. Cajal was a Nobel Prize–

winning scientist and a central figure in the development of modern brain science. Around 1899, he concluded that a particular type of electrically excitable cell was the critical unit of brain function. He called this type of cell a "neuron," after the Greek word for "nerve." His idea became known as "Cajal's neuron doctrine." By 1960, everybody in the field believed it. As with glia, the idea was right there in the name— after Cajal, the study of the brain became known as "neuroscience." Robert Galambos's idea that glial cells had an equally important role to play in making the brain work challenged what every neuroscientist, including Dave Rioch, had believed for their whole careers. It questioned the foundations of the field, risked causing a revolution, and threatened the empire of the neuron. Rioch sensed the risk and tried to shut Galambos down.

Since this confrontation, Galambos's idea has become widely accepted. Scientists do not get fired for having ideas about glia anymore. They are more likely to be promoted. There is an increasing body of evidence that Galambos was right and that glial cells play a vital role in signaling and communication within the brain. They secrete fluids with purposes as yet unknown and may have a crucial influence on brain diseases such as Alzheimer's. One type of glia, star-shaped cells called astrocytes, may be more sensitive signalers than neurons. Fifty years after Galambos's confrontation with Rioch, one scientific review concluded, "Quite possibly the most important roles of glia have yet to be imagined."

The fact that Galambos eventually turned out to be right is beside the point. Organizations are not supposed to work this way. Brilliant, innovative thinking is meant to be encouraged. Galambos and his idea should and could have made a beachhead on a whole new continent of fertile research opportunities. Instead, important discoveries about glia and the brain were delayed for decades. We are learning things today that we could have found out in the 1970s. So why would a distinguished scientist like David Rioch be provoked to anger by an idea proposed by an equally distinguished scientist like Robert Galambos?

The problem was not Rioch. Robert Galambos's story is typical—it happens in almost every organization almost all the time. Kelly Johnson's is not. Both men are examples of what management scholars Larry Downes and Paul Nunes call "truth-tellers":

> Truth-tellers are genuinely passionate about solving big problems. They harangue you with their vision, and as a result they rarely stay in one company for very long. They are not model employees—their true loyalty is to the future, not next quarter's profits. They can tell you what's coming, but not necessarily when or how. Truth-tellers are often eccentric and difficult to manage. They speak a strange language, one that isn't focused on incremental change and polite business-speak. Learning to find them is hard. Learning to understand them, and appreciate their value, is even harder.

Truth-tellers are a bit like the glia of organizations: long overlooked, yet essential for regeneration. They may not be popular. The truth is often awkward and unwelcome, and so are the people who tell it.

As we have seen in our discussions of rejection, confrontations about ideas are hardwired into human nature. The hallmark of a creative organization is that it is much more receptive to new thinking than the world in general. A creative organization does not resent conflicts over concepts; it resolves them. But most organizations are not like Lockheed—they are like Walter Reed. So most truth-tellers are not treated like Kelly Johnson—they are treated like Robert Galambos. We do not walk in a welcoming world when we are given the gift of great thoughts. Great thoughts are great threats.

Kelly Johnson's motto was "Be quick, be quiet, be on time." This was never more important than when he was asked to build *Lulu Belle,* America's first jet fighter. *Lulu Belle* not only flew more quickly than other planes; she was designed and developed more quickly than other planes, too. She had to be: the future of the free world depended on her.

During World War II, planes became faster until they hit a mysterious limit: when they reached 500 miles per hour, they either went out of control or broke apart. Lockheed first experienced the problem in its P-38 Lightning fighter plane, which was so effective that the Germans called it the "fork-tailed devil" and the Japanese called it "two planes, one pilot." Several Lockheed test pilots were killed trying to take the P-38 beyond 500 miles an hour. Tony LeVier, one of Lockheed's greatest test pilots, said that when a plane reached that speed, it felt like "a giant hand shook the plane out of the pilot's control." The problem was so severe that it could not be explored experimentally: at high speeds, model planes were thrown about so forcefully that they could damage a wind tunnel.

As Johnson and his team worked to understand the problem, they uncovered something alarming: the Nazis had already solved it.

On August 27, 1939, four days before World War II began, a plane called the Heinkel He 178 took off from Rostock, on Germany's north coast, and flew over the Baltic Ocean. The He 178 was remarkable because it had no propellers. Instead, it had something no plane before it had ever had: a jet engine.

Planes create waves in the air. The waves travel at the speed of sound. The faster the plane goes, the closer the waves become, until they start to merge. In aerodynamics, this merging is called "compressibility." Compressibility creates a brick wall that planes fly into at around 500 miles an hour—but only if they have propellers.

Jet engines pull air through a funnel. When the air is forced out of the back of the engine, an equal and opposite reaction thrusts the plane forward. Jet planes do not fly into the wall of compressibility; they push off from it. Germany's new jet-powered Messerschmitts, the descendants of the He 178, would be able to outmaneuver, and probably destroy, every other plane in the sky, unless the Allies could develop a jet fighter, too.

Kelly Johnson had wanted to build a jet plane for the U.S. Army Air Forces, the predecessor of the United States Air Force, as soon as he found out about the He 178, but the USAAF told him to make the existing planes fly faster instead. It was only much later, when they discovered Germany's imminent introduction of the jet-powered Messerschmitts, that America's air commanders understood that building a jet plane was the *only* way to make planes fly faster.

The British had developed a jet engine, but attaching it to an existing airplane was ineffective. Jet engines needed whole new aircraft. And so, on June 8, 1943, at 1:30 p.m. exactly, the United States Army Air Forces gave Lockheed a contract to build a jet fighter, and only 180 days in which to do it.

Even Kelly Johnson was not sure he could meet this challenge. Lockheed was already building twenty-eight planes a day, working three shifts every day except Sunday, when it did one or two. The company had no extra engineering capacity, no extra space, and its equipment was in constant use. Lockheed's president, Robert Gross, told Johnson, "You brought this on yourself, Kelly. Go ahead and do it. But you've got to rake up your own engineering department and your own production people and figure out where to put this project."

These apparently impossible constraints gave us the model creative organization.

Johnson believed that engineers should be as close to the action as possible, so he used Lockheed's lack of spare capacity as an excuse to build a "lean" organization, where the muscles of his team—the designers, engineers, and mechanics—had direct connections to one another, without the fat of managers and administrative staff keeping them apart.

The lack of extra space, as well as the need for high security, gave him an excuse to build an isolated, insulated organization. No one else was allowed into the Skunk Works' box-and-tent "building." This was not just to keep the project covert. It had another benefit: shared secrets and an exclusive workspace gave the team a unique bond.

Inside the tent, a "scoreboard calendar" counted down the 180 days, keeping everyone focused on creation's most precious resource: time.

The challenges grew toward the end of the project: half the team fell ill because of the workload, the barely heated makeshift building, and colder-than-usual midwinter weather. They had to build the plane without ever seeing its engine—it had been shipped from Britain, but the expert sent with it in secret was arrested on suspicion of spying because he could not explain why he was in America. Then the day before the plane was scheduled to fly, the engine exploded. There was no choice but to wait for another—the only other one in existence.

The proof of the organization was the result. Despite these obstacles, the Skunk Works beat the schedule by thirty-seven days, and *Lulu Belle* flew the first time.

5 | THE SECRET OF BERT AND ERNIE

Mike Oznowicz and his wife, Frances, escaped the Nazis twice during the 1930s. First they fled from Holland to North Africa and then, when the war followed them, they fled from North Africa to England. They had two children while they were in England; the second, Frank, was born in the barracks town of Hereford in May 1944. In 1951, Mike and Frances spent their last dollars moving their family to the United

States, eventually settling in California, where Mike found work dressing windows.

Mike and Frances's passion was puppets. Both were active members of the Puppeteers of America, a nonprofit founded in 1937 to help promote and improve the art of puppetry. In 1960, the Puppeteers' annual Puppetry Festival was held in Detroit, Michigan. Mike and Frances befriended a first-time attendee named Jim Henson. Henson, his wife, and their three-month-old daughter had driven five hundred miles from their home in Bethesda, Maryland, in a Rolls-Royce Silver Shadow to attend the show. One day a friend drove the Rolls-Royce around Detroit while Henson performed puppetry through the sunroof with a hand-and-rod frog he called "Kermit."

Henson became close to the Oznowiczs. In 1961, when the Puppetry Festival was held in Pacific Grove, California, they introduced him to their son Frank, who had just turned seventeen. Frank was a skilled manipulator of the puppets on strings called marionettes, and he won the festival's talent contest, even though he preferred baseball and, he said, practiced puppetry only because he was the child of a family of puppeteers.

Henson was building a successful business making TV commercials with a new style of puppets he called "Muppets." Henson thought Frank had great talent and wanted to hire him. At first, Frank declined—he wanted to be a journalist, not a puppeteer, and he was only seventeen years old. But there was something about Jim Henson, and the meeting, that he could not forget: "Jim was this very quiet, shy guy who did these absolutely fucking amazing puppets that were totally brand new and fresh, that had never been done before."

After Frank finished high school, he agreed to take a part-time position with Henson's company, Muppets, Inc., and also enrolled in City College in New York so he could get an education. But within two semesters, Frank stopped going to college and started working full-time with Henson. Frank said, "What was going on with the Muppets was too exciting."

By 1963, when Frank joined Henson, the Muppets were moving beyond commercials. A popular country music singer named Jimmy Dean was planning a variety show for ABC Television, and he wanted Henson to provide a puppet for the show. Henson created Rowlf, a brown, floppy-eared dog. Rowlf got up to eight minutes of airtime per episode, often upstaged Dean, and received thousands of letters from fans each week.

Rowlf was what is known as a "live-hand Muppet." Some Muppets, like Kermit the Frog, are "hand-and-rod Muppets": a single puppeteer—technically, a "Muppeteer"—puts one hand in the Muppet's head and uses the other to manipulate the Muppet's hands using rods. Live-hand Muppets need *two* Muppeteers: one Muppeteer puts one hand, usually the right one, into the Muppet's head and the other hand into the Muppet's glovelike left hand. A second Muppeteer puts his or her right hand into the Muppet's right hand. The two Muppeteers stand close together and must think and move as one. Henson was Rowlf's voice, head, mouth, and left hand; Frank was Rowlf's right hand. One night, Jimmy Dean, the show's host, stumbled saying "Oznowicz" on air and accidentally gave Frank a more magical name: Oz.

Oz and Henson were at the beginning of what would become a potent creative partnership.

A few years later—partly because of Rowlf—Henson, Oz, and the rest of Muppets, Inc., were recruited to work on a new television series for children to be called *Sesame Street*.

While they were preparing for the first show, Henson and Oz found two new Muppets in the rehearsal room, made and designed by master Muppet maker Don Sahlin, the creator of Rowlf. One was a tall hand-and-rod puppet with a long yellow head like a football about to be kicked, crossed by one thick eyebrow. The other was its opposite, a short live-hand puppet with a squat orange head, no eyebrows, and a meadow of black hair.

Henson took the yellow puppet and Oz took the orange one, trying to discover the characters inhabiting them. The puppets felt wrong. They switched. Henson took the short orange guy with the scared-cat hair; Oz took the tall guy with the unibrow. Everything clicked. The yellow Muppet, played by Oz, became "Bert," careful, serious, and sensible; the orange Muppet, played by Henson, became "Ernie," a playful, funny risk taker. Bert was the kind of guy who wanted to be a journalist, not a puppeteer. Ernie was the kind of guy who would cruise Detroit in a Rolls-Royce waving a frog through the sunroof. And yet, somehow, Bert and Ernie were kindred spirits, worth more together than apart.

The first episode of *Sesame Street* was broadcast on Monday, November 10, 1969. After the words "In Color," two clay animation monsters appear, followed by an archway with the words "Sesame Street." The monsters walk through the arch, the screen fades to black, and the show's theme song, "Can You Tell Me How to Get to Sesame Street?" begins, sung by a choir of children over clips of real urban kids—not the scrubbed and tailored angels normally seen on television at that time—playing in city parks. The title sequence ends, and the show opens on the green street sign that says, "Sesame Street," while an instrumental version of the song, played on harmonica by jazz musician Toots Theilemans, begins. A black schoolteacher named Gordon is showing a little white girl named Sally around the neighborhood. After introducing her to some human characters and an eight-foot-tall full-costume Muppet called "Big Bird," Gordon hears singing coming from the basement of 123 Sesame Street and points to the basement window, saying, "That's Ernie. Ernie lives down in the basement, and he lives there with his friend Bert. Whenever you hear Ernie singing, you can bet he's taking a bath."

The show cuts to Ernie in the bathtub, singing while scrubbing.

ERNIE: Hey, Bert. Can I have a bar of soap?

BERT (*entering*): Yah.

ERNIE: Just toss it into Rosie here.

BERT (*looking around, perplexed*): Who's Rosie?

ERNIE: My bathtub. I call my bathtub Rosie.

BERT: Ernie, why do you call your bathtub Rosie?

ERNIE: What's that?

BERT: I said, why do you call your bathtub Rosie?

ERNIE: Because every time I take a bath, I leave a ring
around Rosie.

Ernie giggles a glottal, staccato giggle. Bert looks at the camera, as if asking the audience whether they can believe this guy. With that sequence, Bert and Ernie became the first puppets to appear on *Sesame Street*. They are still major characters today.

Bert and Ernie's close relationship has often aroused suspicion. What were two male puppets doing together? Why were they so close? Pentecostal pastor Joseph Chambers of Charlotte, North Carolina, thought he knew: "Bert and Ernie are two grown men sharing a house and a bedroom. They share clothes, eat and cook together, and have blatantly effeminate characteristics. In one show Bert teaches Ernie how to sew. In another they tend plants together. If this isn't meant to represent a homosexual union, I can't imagine what it's supposed to represent."

But no, Bert and Ernie are not gay. To discover what the characters represent, we need look no further than the men inside the puppets. Not only were Henson and Oz Bert and Ernie, Bert and Ernie were Henson and Oz. *Sesame Street* writer Jon Stone recalls, "Their relationship reflected the real-life Jim-Frank relationship. Jim was the instigator, the teaser, the cutup. Frank was the conservative, careful victim. But essential to the rapport was the affection and respect which these two men held for each other. Ernie and Bert are best friends; so it was with Jim and Frank."

Some of the greatest creative work comes from people working in twos. The partnership is the most basic unit of creative organization, and it holds many lessons for how to build creative teams. Some creative partners are married, like Pierre and Marie Curie; some are family, like Orville and Wilbur Wright; but most are neither. They may not even be friends. They are people like Simon and Garfunkel, Warren and Marshall, Abbott and Costello, Lennon and McCartney, Page and Brin, Hanna and Barbera, Wozniak and Jobs, Henson and Oz.

As in the story of Bert and Ernie, the intimacy of the creative partnership confuses some people, perhaps because they overestimate the importance of individuals.

The secret of Bert and Ernie is that nothing is created alone. Steve Wozniak's advice to "work alone," mentioned earlier, is not as simple as it seems. As Robert Merton observed, we never act as individuals without interacting with myriad others—by reading their words, remembering their lessons, and using tools they made, at the very least. A partnership puts this interaction into the same room.

6 | WHEN THE ROAD SEEMS LONG

In a creative partnership, the alternating nature of ordinary conversation and the problem-solution loops of ordinary thinking combine: partners use the same creative process as individuals but do their thinking aloud, seeing problems in each other's solutions, and finding solutions to each other's problems.

Trey Parker and Matt Stone have been creative partners since they met at the University of Colorado in 1989. In 2011, they won nine Tony awards for *The Book of Mormon,* a Broadway musical they co-wrote with

Robert Lopez; they have created movies, books, and video games; and they are best known for *South Park,* an animated television series they created in 1997. Parker and Stone have written, produced, and voiced hundreds of episodes of *South Park,* most of them made, from conception to completion, in six days.

The process begins in a conference room in Los Angeles on a Thursday morning, where Parker and Stone discuss ideas with their head writers and start to create the show that will air the following Wednesday. Stone describes the room as "a safe place, because for all the good ideas that we get, there's a hundred not so good ones." No one else is allowed in, but, in 2011, Parker and Stone let filmmaker Arthur Bradford put remote cameras in the room to make a documentary called *6 Days to Air: The Making of South Park.*

On day one of the film, Parker and Stone discuss script ideas, including the Japanese tsunami, bad movie trailers, and college basketball, ad-libbing possible scripts as they go, much like Henson and Oz trying out Bert and Ernie for the first time. By the end of the day, Parker and Stone have nothing—or, at least, nothing they like. Parker is worried. He tells Bradford, "There's a show on this Wednesday. We don't even know what it is. Even though that's the way we've always done it, there's a little voice saying, 'Oh, you're screwed.'"

On the morning of day two, Parker makes a suggestion to Stone: "Let's try this. Let's go to eleven-thirty trying to come up with something completely new; then from eleven-thirty to twelve-thirty we'll pick which of yesterday's ideas we're going to do."

Stone is skeptical: "A whole other show?"

But rather than argue for improving the existing ideas, Stone tries the process. Eventually, Parker throws out an idea about something he finds frustrating: "Last night, I went onto iTunes, and that window came up again that says, 'Your iTunes is out of date,' you know, which happens every time. 'God damn it. Here it goes again. I got to download another version of iTunes.' How many times have I hit 'Agree' to those terms and conditions, and I've never even read one line of them?"

Stone laughs, then suggests how Parker's frustration with iTunes might yield a plot: "The joke is that everyone always reads the terms and conditions except for Kyle." (Kyle is one of the show's main characters.)

Stone, speaking later, explained what happened next: "And then we said, 'Oh wait, this is actually starting to be something.'"

The pattern—Parker directing the process and finding the points of departure, and Stone refining and building on them—is typical of Parker and Stone's working relationship. Partnerships tend not to be hierarchical, in the sense that one person has authority over the other, but they are seldom leaderless. In Parker and Stone's partnership, Parker leads. Stone says, "Even though we're a partnership, and we each bring something different to the table, the way that the stories are expressed is completely through Trey. It's like Trey's the chef. Whatever I've got channels through him."

Parker agrees, but he is under no illusions about Stone's equal importance. Referring to another famous partnership from the rock band Van Halen, Parker says, "You can sit there and say, 'Well, it's all Eddie Van Halen,' but as soon as David Lee Roth leaves, you say, 'Well, forget that band.' Eddie can sit there and say, 'I write everything,' but you're not Van Halen without David Lee Roth."

On Monday, with less than three days until airtime, the script is still not complete. Animation and voice work is under way on the main plot, which features Kyle being forced to do the crazy things he agreed to when he accepted the iTunes terms and conditions. But the show lacks a subplot and an ending. Parker begins the day by describing the remaining problems to Stone: "We are in danger of doing our typical first-show thing where we've just got way too many ingredients; we haven't introduced the idea of apps at all; and I am worried about time—whatever this thing is at the end, it is going to have to be fast."

Parker then starts solving these problems—by, for example, describing a subplot where another main character, Eric, tries to persuade his mother to buy him an iPad. Stone's role here is evaluation: he laughs as Parker acts out the idea.

Parker is worried. That evening, he tells documentary maker Bradford, "I am pretty scared right now because I am up to twenty-eight pages of script and I still have five scenes to write. Each scene's about a minute long usually, so this is going to end up being about a forty-page script, which just becomes brutal, because I have to start taking scenes and figuring out how to do this same thing in half the time."

Even in a partnership, the literal, physical act of writing—choosing the words, rather than having the ideas—is an individual activity. This is what Wozniak means when he says, "Work alone." Two people and a blank page is no formula for creation: a pencil is a one-person device. Stone does not hover on Parker's shoulder trying to be helpful. He is in another room, working on script edits. Parker says, "I hate writing because it is so lonely and sad. I know everyone's waiting for me to get it done, and it is a battle of fighting over lines and trying to figure out what the best way to say things is. I just hate it so much."

As Monday ends, Parker paces while Stone watches from a couch. Both men are rubbing their heads and plucking at the bridges of their noses. Parker summarizes the current problem: "We're a minute short and I have four scenes to write."

In four days, he has gone from worrying about having no material to worrying about having too much. Stone says he hits rock bottom every Sunday and Parker hits rock bottom a day later. Sure enough, on Monday, Parker says, "I feel terrible about the episode. I am embarrassed we are putting this piece of shit on the air."

The laughter of a few days ago is gone. Parker and Stone stalk the studio, hunched and miserable.

Tuesday, the day before the show will air, starts with exhausted animators sleeping under their desks or at their keyboards. At six a.m., while the sun rises, Parker and Stone meet alone in the writers' room. Parker has erased the crude pictures that were on the whiteboard a few days ago; now there is a flat list of scenes with names like "Playground," "Eric's Home," "Jail Scene," and "At the Genius Bar." Parker stands at the whiteboard with a marker while Stone leans back in an armchair, his hands behind his head.

The roles have changed. Parker is no longer leading. He is pitching ideas: "Beginning of act two, we come back, and that's when it's 'Okay, the Geniuses are going to see us now.' And then, act three: we just start with the unveiling of the thing. And then we go to, they're doing the bubble thing, and Gerald flips out, joins Apple. We're back, and that's it."

And Stone is not being led; he is coaching and cajoling. He sounds paternal when he says, "That's great. Yeah, that actually works."

Reinvigorated, Parker returns to the keyboard. An hour later, the animators are being woken up and handed the completed script for *South Park*'s episode 1501: the first of the fifteenth season and the 211th Parker and Stone have written.

The story shows how many creative partnerships work. Parker trusts Stone. Stone complements Parker. Parker may appear to make a greater creative contribution, but Stone enables it, in particular by giving Parker emotional support during the loneliness and stress of creation. Stone creates, too, and Parker enables *that* by providing impetus. Partners create together by helping each other create individually.

7 | THE WRONG TYPE OF ORGANIZATION

The common thread that connects one person creating to two people creating can—or should—extend to larger groups of people, too. Creative partners talking sound a lot like creative individuals thinking aloud, and nothing needs to change when the group gets larger. The purpose of a creative conversation is to identify and solve creative problems, such as "What should this episode be about?" or "What order should these scenes be in?" The only participants in the conversation should be people who can make a contribution to answering these questions, which is why Parker and Stone's writers' room is a "safe place," off-limits to all but a few writers. There is no room for managers, "devil's advocates," or any other species of spectator in a creative conversation. This conversation is the main purpose of creating in a group. The detailed creative work is still done alone, unless help—practical, emotional, or both—is needed to get past the inevitable pressures and failures.

Parker and Stone's company, South Park Digital Studios, is a lot like Lockheed's Skunk Works: it is part of a big corporation, Viacom; it is isolated in its own physical location; and it is capable of working almost impossibly fast. While it takes six days to make an episode of *South Park*, it takes six *months* for most other production companies to make an episode of most animated TV shows.

In the wrong type of organization, Parker and Stone's creative talent can quickly become destructive. In 1998, Viacom asked the two men to make a *South Park* movie with another one of its subsidiaries, Paramount Pictures. Parker and Stone started fighting with the Paramount executives almost as soon as production began. One of their first battles was about the movie's rating. Parker and Stone wanted a movie with themes and language that would make it R-rated—meaning that no one under seventeen would be admitted unless they were accompanied by a parent or adult guardian. Paramount wanted a PG-13 rating—a milder movie that anyone could see, although parents would be warned that some content may be inappropriate for children under thirteen.

Parker rebelled: "After they showed us graphs of how much more money we'd make with a PG-13, we were like, 'R or nothing.'"

Parker declared what he later called "war." Paramount sent them tapes of trailers for the movie; Parker and Stone broke them in half and mailed them back. They sent rude faxes to everyone they knew at Paramount, including one, titled "A Formula for Success," that said, "Cooperation + you doing nothing = success." Parker stole the only copy of a censored promotional videotape to stop it from being broadcast on MTV. After this incident—the tape was the result of several days and nights of hard work by Paramount employees—Paramount threatened to sue Parker and Stone.

Parker and Stone's biggest protest against Paramount was the movie itself. They turned the film into a full-length musical about their frustration with Paramount's attempts to censor them. In *South Park: Bigger, Longer & Uncut,* the United States declares war on Canada because of a Canadian TV show with bad language; a schoolteacher tries to rehabilitate cussing children by singing a song based on *The Sound of Music*'s "Do-Re-Mi"; and characters say things like "This movie has naughty language, and it might make you kids start using bad words," and "I'm sorry! I can't help it!! That movie has warped my fragile little mind."

By the short-term standards of profit and loss, the Paramount–South Park collaboration was a success: the film grossed $83 million against a budget of $21 million, won awards, and Parker and co-writer Marc Shaiman received an Oscar nomination for their song "Blame Canada." But from the longer-term perspective of building a creative organization, the project was a catastrophe, and an expensive one: despite the positive outcome, and despite owning the rights to sequels, Paramount will never be able to make another *South Park* movie.

Parker told *Playboy,* "They couldn't pay us enough to work with them again."

Stone added, "You had marketing battles, legal battles, all these battles. Even with the clout of having this huge franchise that had earned Viacom hundreds of millions of dollars, the studio did everything they could to beat us down and beat the spirit out of the movie."

If Parker and Stone sound childish, it is because they *are* childish—in the best possible way. The social skills that enable creation through cooperation—and the antisocial behavior that can result when creation is excessively controlled—are things we all have as children but that are educated out of most of us as we grow up. We develop our ability to create in groups when we develop our ability to talk, but we often lose it during our school years, and we may have lost it completely by the time we start our first job. One of the first people to discover this was a man in Belarus in the 1920s. One of the best ways to demonstrate it is with a marshmallow.

8 | A LITTLE LESS CONVERSATION

In 2006, Peter Skillman, an industrial designer, gave a three-minute presentation at a conference in Monterey, California. He spoke immediately after former Vice President and future Nobel Prize winner Al Gore and immediately before spacecraft designer Burt Rutan. Despite the lack of time and the difficult billing, Skillman's talk made a big impact. It described what he called "the marshmallow challenge," a team-building activity he developed with Dennis Boyle, a founding member of the design consultancy IDEO. The challenge is simple. Each team is given a brown paper bag containing twenty sticks of uncooked spaghetti, a yard of string, a yard of masking tape, and a marshmallow. The goal is to build the tallest possible freestanding structure that can take the weight of the marshmallow. The team members cannot use the paper bag, and they cannot mess with the marshmallow—for example, they cannot make it lighter by eating some of it—but they can break up the spaghetti, string, and masking tape. They have eighteen minutes, and they cannot be holding their structure when the time is up.

Skillman's most surprising finding: the best performers are children aged five and six. Skillman says, "Kindergartners, on every objective measure, have the highest average score of any group that I've ever tested." Creative professional Tom Wujec confirmed this: he conducted marshmallow challenge workshops more than seventy times between 2006 and 2010 and recorded the results. Kindergartners' towers average twenty-seven inches high. CEOs can only manage twenty-one-inch towers, lawyers build fifteen-inch towers, and the worst scores come from business school students: their towers are typically ten inches high, about one-third the height of the towers built by kindergartners. CEOs, lawyers, and business school students waste minutes on power struggles and planning, leave themselves only enough time to build one tower, and do not uncover the hidden assumption that makes the challenge so challenging: marshmallows are heavier than they look. When they finally figure this out, they have no time left to do anything about it. Wujec recounts those last moments: "Several teams will have the powerful desire to hold on to their structure at the end, usually because the marshmallow, which they just placed onto their structure moments before, is causing the structure to buckle."

Young children win because they collaborate spontaneously. They build towers early and often rather than wasting time fighting for leadership and dominance, they do not sit around talking—or "planning"— before they act, and they discover the problem of the marshmallow's weight quickly, when they have lots of time left to solve it.

Why do children do this? That question is answered by the work of Lev Vygotsky, a psychologist from Belarus. In the 1920s, Vygotsky discovered that the development of language and creative ability are so connected they may even be the same thing.

The first thing we do with speech is organize our surroundings. We name important people, like "Mama" and "Dada," and we name important objects, both natural, like "dog" and "cat," and man-made, like "car" and "cup." The second thing we do with speech is organize our behavior. We can set ourselves goals, like chase the dog or grab the cup, and communicate needs, like ask for Mama. We may have had these goals and needs before we could speak, but words allow us to make them more explicit, both to ourselves and to others. When we know the word for dog, we are more likely to chase a dog, because we are more capable of *deciding* to do so. This is why young children chasing dogs can often be heard saying "dog" to themselves again and again. Words beget wishes. The next thing we do with speech is create: when we can manipulate a word, we can manipulate the world. Or, as Vygotsky said:

> Although children's use of tools during their preverbal period is comparable to that of apes, as soon as speech and the use of signs are incorporated into any action, the action becomes transformed and organized along entirely new lines. The specifically human use of tools is thus realized, going beyond the more limited use of tools possible among the higher animals.

For example, when Vygotsky's research associate Roza Levina asked Milya, a four-year-old girl, to draw a picture of the sentence "The teacher is angry," Milya was unable to complete the task. Levina reports what Milya said:

> "The teacher is angry. I can't draw the teacher. This is how she looks." (She draws, pressing hard with a pencil.) "It is broken. It is broken, the pencil. And Olya has a pencil and a pen." (Child fidgets on her chair.)

Milya's response is typical of a child in the first stage of using language—labeling her world. Her speech is not yet a system of signs that helps her achieve goals; it is a narration of the here and now.

Anya, three years and seven months old, is younger than Milya, but she is in the next stage of development. (Another of Vygotsky's discoveries was something we now take for granted: that children's minds develop at different speeds.) Vygotsky put some candy on top of a cupboard, hung a stick on the wall, and asked Anya to get the candy. At first there was a long silence. Then Anya started talking about, and *working on,* the problem. Vygotsky reports:

"It's very high." (She climbs up onto the divan and reaches for the candy.) "It's very high." (She reaches.) "You can't get it. It's very high." (She grasps the stick and leans on it, but she does not use it.) "I can't get it. It's very high." (She holds the stick in one hand and reaches for the candy with the other.) "My arm's tired. You can't get it. We have a tall cupboard. Papa puts things up there, and I can't get them." (She reaches.) "No, I can't reach it with my hand. I'm still little." (She stands up on a chair.) "There we go. I can get it better from the chair." (She reaches. She stands on the chair, and swings the stick. She takes aim at the candy.) "Uh-uh." (She laughs and pushes the stick forward. She glances at the candy, smiles, and gets it with the stick.) "There, I got it with the stick. I'll take it home and give it to my cat."

The difference between Anya and Milya is one of development, not ability. Milya will soon be able to do what Anya did: use language not just to label the world but also to manipulate it in pursuit of a goal. Vygotsky did not have to ask Anya to think aloud about reaching the candy—children at her stage do that anyway. Anya's thoughts connect to her actions because we do not manipulate the world, then describe what we did afterward. We manipulate language so we can manipulate the world.

Language and creation are so interconnected that you cannot have one without the other. Language, in this sense, means a system of symbols and rules that allows us to make and manipulate a mental representation of past, present, and possible future states. People who prefer pictures to words, for example, still move symbols around—

some of the symbols just happen to be images. Anya developed this ability relatively early; children normally move from labeling with language to manipulating with language between the ages of four and five.

The connection between language and creating has an important consequence: once children can solve problems by talking about what they are doing, they have the basic skills they need to create with others.

The surprising thing about the marshmallow challenge, then, is not the performance of the children but the performance of the adults. The business students who build a ten-inch tower would have built a twenty-seven-inch tower when they were in kindergarten. Where did those extra seventeen inches go? What happened to the students in the intervening years?

The business students, like most of the rest of us, lost a lot of their capacity to cooperate. The focus on individual accomplishment in their education and environment taught them that it was more valuable to perform individual tasks, especially solving problems with definite answers, than to work on ambiguous things in teams. The natural collaborative ability they developed as children got squashed like their marshmallow towers.

Even worse, by the time children become adults, they have learned that talking is an alternative to doing. At school, most work is done individually and quietly—especially most of the work that gets graded. One of the most common classroom rules is "No talking." The message is clear: you cannot do and talk at the same time.

This division between words and actions persists into the workplace, where groups solve problems by talking—or "planning"—until they agree on what they think is the one best answer, then take action. Children do not hold meetings at school; they discover them as adults, at work. Children see the marshmallow challenge as a chance to collaborate; adults treat it like a meeting. All the children in a team build and experiment, compare results, learn from one another, and create as a community as soon as the clock starts ticking. They do not discuss this in advance. They just get on with it. All the adults in a team do nothing for the first few minutes, because they are talking instead; then most of them do nothing but watch—or "manage"—someone else building a tower for the remainder of the time. According to Tom Wujec's data, kindergartners try putting the marshmallow on the tower an average of five times during the eighteen minutes. Their first attempt usually happens between the fourth and fifth minutes. Business students typically put the marshmallow on the tower once, at the eighteenth—or last—minute.

Vygotsky's research explains why children act when adults plan. The connection between expression and action is stronger when we are younger. This is most obvious in experiments that involve choice. Vygotsky asked four- and five-year-old children to press one of five keys that corresponded to a picture they were shown. The children thought not with words but with actions. Vygotsky notes:

> Perhaps the most remarkable result is that the entire process of selection by the child is *external,* and concentrated in the motor sphere. The child does her selecting while carrying out whatever movements the choice requires. Adults make a preliminary decision internally and subsequently carry out the choice in the form of a single movement that executes the plan. The child's movements are replete with diffuse gropings that interrupt and succeed one another. A mere glance at the chart tracing the child's movements is sufficient to convince one of the basic motor nature of the process.

Or: adults think before acting; children think *by* acting.

Talking while acting is useful, but talking *about* acting is *not*—or, at least, not often, and not for long. This is why "Show me" is such a powerful thing to say. "Show me" stops speculation and starts action.

Another thing adults have learned that kindergartners have not is that groups must be hierarchical. Adults start with some team members locking horns for leadership. Children start with everyone working together.

Creative partnerships are barely hierarchical—they would not be "partnerships" if they were—so little or no energy is expended on dominance rituals. Jim Henson was senior to Frank Oz in every way but one: when Henson and Oz created together, they were equals. There is no partnership without equality. Henson and Oz did not waste their time on power struggles; they spent it all on doing, talking aloud like the children in Vygotsky's research, solving problems, and helping each other grow. The birth of Bert and Ernie is a perfect example. Henson and Oz did not hold a meeting or make plans. They picked up the puppets and thought aloud until Bert and Ernie appeared.

9 | WHAT ORGANIZATIONS ARE MADE OF

In 1954, something unprecedented happened at six trials in the U.S. Courthouse in Wichita, Kansas. These were typical trials with typical cases, typical defendants, and typical convictions and acquittals. Only the heating units in the jury room were strange. They contained hidden microphones, put there by University of Chicago researchers, who used them to record the jurors' deliberations. The judge and lawyers knew about the microphones, but the jurors did not.

The recordings were sealed until a final judgment was entered in each case and all appeals were dismissed. Then the researchers analyzed the interactions to learn about group behavior in a jury room. When the findings were published a year later, they caused a nationwide scandal. In one of the first privacy controversies, the Senate Subcommittee on Internal Security subpoenaed the researchers, and more than a hundred newspaper editorials condemned them for threatening the foundation of the American legal system.

The scandal has been forgotten, but the method has not. Harold Garfinkel, one of the researchers who analyzed the jury room tapes, called it "microsociology." Scientists have now conducted thousands of experiments using microphones and video cameras to understand the human behavioral minutiae that compose society.

One reason traditional sociology, or "macrosociology," looks at large groups from afar over long periods of time is technological. When the social sciences were first conceived—in large part by Frenchman Émile Durkheim in the 1890s—there was no practical way to record and observe everyday interactions in detail. Microsociology became possible only in the 1950s, with the invention of the magnetic tape recorder, the transistor, and mass-production electrical microphones.

Like traditional sociologists, business writers—often, by the way, former business school students—typically look at organizations as if they were flying high above them. They see the big picture—the mergers, changes in stock price, and major product launches that are the equivalent of the freeways, neighborhoods, and parks we see through the window of a descending airplane—but, with the exception of a few senior executives, the individuals are invisible.

You cannot learn much by looking at an organization from the sky. Organizations exist only on the ground. They are not, as is commonly claimed, made of people. Organizations are made of people *interacting.* What an organization organizes is everyday human interactions.

Microsociology shows us that these interactions are not trivial. Everything that happens between two or more people is rich in meaning.

Before microsociology, the dominant assumption was that people in groups made decisions using reasoning, in a series of steps something like this:

1. Define the situation.
2. Define the decision to be made.
3. Identify the important criteria.
4. Consider all possible solutions.
5. Calculate the consequences of these solutions versus the criteria.
6. Choose the best option.

Microsociology showed conclusively that we seldom think this way, especially not in groups. In group interactions, our decisions are more likely to be based on unwritten rules and cultural assumptions than on pure reason. Ludwig Wittgenstein, an Austrian British philosopher, said that these interactions, which on the surface look like nothing more than talking, are like a game, because they consist of "moves" and "turns." He called the game *Sprachspiel,* or "the language game."

In a group, words are heard in a context that includes emotion, power, and existing relationships with other group members. We are all social chameleons, adjusting our skin to blend in with, or sometimes stand out from, whatever crowd we happen to be in.

Sociologist Erving Goffman called the moves in the language game "interaction rituals." Later, his colleague Randall Collins called series of these moves "interaction ritual chains." The chain starts with the situation—for example, a business meeting. The way each individual behaves in the meeting will depend on a number of things: their level of authority, their mood, their previous experience in similar meetings, and their current relationships with the other people in the room. All these things change their behavior. They will not act the way they might in a different situation—for example, when they are unwell and visiting the doctor. In the meeting, the greeting "How are you?" signifies nothing more than courtesy. Collins writes, "'How are you?' is not a request for information, and it is a violation of its spirit to reply as if the interlocutor wanted to know details about one's health."

In the other case, when a person is visiting the doctor, the question "How are you?" at the start of the appointment *is* a request for information. It would be a violation *not* to provide details about one's health. The same person being asked the same question gives a different reply because they are participating in a different ritual.

Organizations are made of rituals—millions of small, moments-long transactions between individuals within groups—and it is these rituals that determine how much an organization creates.

10 | RITUALS OF DOING

The biggest lesson from the story of Kelly Johnson and the Skunk Works is that creation is doing, not saying. The most creative organizations prioritize rituals of doing; the least creative organizations prioritize rituals of saying, the most common of which is the meeting. "Meeting" is a euphemism for "talking"; therefore, meetings are an alternative to work. Despite this, the average office worker attends six hour-long meetings a week, almost a full working day. If an organization uses Microsoft's Outlook software to automatically schedule meetings, their employees attend even more meetings—nine hour-long meetings a week. There is no creating in meetings. Creation is action, not conversation. Creative organizations have external meetings—for example, with customers, as Lockheed did to win its wartime contracts to make planes—but the more creative an organization is, the fewer *internal* meetings it tends to have, and the fewer people tend to be at those meetings. The result is more people spending more time at the coal face of creation.

Much of what happens in internal meetings is called "planning," but planning is of limited value, because nothing ever goes according to plan. Kelly Johnson had little use for plans and did not need to know the details of how things were going to happen before doing them.

Engineering plans are important for getting a product built, but engineering plans are doing, not saying. Even then, some engineering plans are made after the product is built. Johnson describes his first day at Lockheed:

> I was assigned to work with Bill Mylan in the tooling department, designing tools for assembly of the Electra. Mylan was an old hand and knew his business. "I'll build them, kid, and you can draw them later," he explained to me.

You cannot control the future. Being too rigid about making things happen the way you planned stops you from reacting to emerging problems and causes you to miss unexpected opportunities. Have high expectations about what and few expectations about how. This is the opposite of the way most organizations operate. Many "executives" spend half of their week in "planning" meetings and the other half preparing for them. You cannot build a plan that predicts your setbacks—like the engine expert being arrested as a spy, or his engine exploding the first time you turn it on—but you can build an organization that executes anyway.

Saying instead of doing is worse than unproductive: it is counterproductive. In 1966, Philip Jackson, one of the psychologists who discovered that teachers do not like creative children, introduced a new term to describe how organizations transmit values and shape behavior: the "hidden curriculum."

Jackson used the term to describe schools:

> The crowds, the praise, and the power that combine to give a distinctive flavor to classroom life collectively form a hidden curriculum, which each student (and teacher) must master if he is to make his way satisfactorily through the school. The demands created by these features of classroom life may be contrasted with

the academic demands—the "official" curriculum, so to speak—to which educators traditionally have paid the most attention.

We learn the hidden curriculum as children, when our minds are eager, we hunger for friends, and we are most afraid of shame. We learn it without knowing: the hidden curriculum is a set of unwritten rules, implied, often at odds with what we are told. We learn the opposite of the official curriculum: that originality ostracizes, imagination isolates, risk is ridiculed. You faced a choice as a child you may not remember: to be yourself and be alone or to be like others and be with others. Education is homogenization. This is why nerds are targets and friends move in herds.

We carry this lesson through life. Education may be forgotten, but experience gets ingrained. What we divide into discrete periods like "high school," "college," and "work" is in fact a continuum. And so the hidden curriculum operates in all organizations, from corporations to nations. Jackson says:

> As institutional settings multiply and become for more and more people the areas in which a significant portion of their life is enacted, we will need to know much more than we do at present about how to achieve a reasonable synthesis between the forces that drive a person to seek individual expression and those that drive him to comply with the wishes of others.

Organizations are a competition between compliance and creation. The leaders of our organizations may ask us to create sometimes, but they demand that we comply always. Compliance is more important than creation in most organizations, no matter how much they pretend otherwise. If you comply but do not create, you may be promoted. If you create but do not comply, you will be fired. When rewards are given for compliance, not contribution, we call it "office politics." We are required to comply not with what the organization *says*, but with what the organization *does*. If a CEO stands up and gives an annual all-company PowerPoint presentation about his love of innovators and risk takers, then allocates most of his company's money to the old product groups and gives all his promotions to the people who manage them, he sends a clear signal to anybody who understands the hidden curriculum: do what the CEO does, not what the CEO says. *Talk about* innovating and taking risks, but do not *do* it. Work in the old product groups and focus your actions on old products. Leave the innovative, risky products to less organizationally adept, more creative people, who will be fired as soon as they fail and who will fail because they are not given any resources. This approach to getting ahead is one many organizations demand, although they do not realize it and will not admit it. Jackson writes:

> No matter what the demand or the personal resources of the
> person facing it there is at least one strategy open to all. This is
> the strategy of psychological withdrawal, of gradually reducing
> personal concern and involvement to a point where neither the
> demand nor one's success or failure in coping with it is sharply felt.

Can someone both be inventive and follow the hidden curriculum that puts compliance and loyalty over creation and discovery? Perhaps, but the two things are opposites:

> The personal qualities that play a role in intellectual mastery are
> very different from those that characterize the Company Man.
> Curiosity, as an instance, is of little value in responding to the
> demands of conformity. The curious person typically engages in a
> kind of probing, poking, and exploring that is almost antithetical
> to the attitude of the passive conformist. Intellectual mastery calls

for sublimated forms of aggression rather than for submission to constraints.

Also, why bother? Why spend the energy and imagination needed to maintain a false identity—to be a conforming Clark Kent so you can keep your creative superself hidden—when you can get equally good results by conforming without creating or by taking your creative abilities somewhere where they will be appreciated? This is the dilemma creative people face everywhere. They seldom choose to resolve it by being secretly creative. Most people resign or become resigned after taking a new idea to their boss for evaluation. Proposing something new is a high-risk transaction. For Kelly Johnson at Lockheed in the 1930s, it worked. For Robert Galambos at the Walter Reed Army Institute of Research in the 1960s, as in most organizations most of the time, it did not.

Building a creative organization is hard, but keeping it creative is many times harder. Why? Because every paradigm changes, and only the best creators can change with the consequences of their creations.

In the summer of 1975, a few months after the fall of Saigon and the end of the Vietnam War, a Skunk Works engineer named Ben Rich presented an idea to Kelly Johnson. It was a design for an airplane shaped like an arrowhead: flat, triangular, and sharply pointed. Rich and his team called it the "Hopeless Diamond." Johnson's initial reaction was not positive. Rich said, "He took one look at the sketch of the Hopeless Diamond and charged into my office. Kelly kicked me in the butt—hard too. Then he crumpled up the proposal and threw it at my feet. 'Ben Rich, you dumb shit,' he stormed, 'have you lost your goddam mind?'"

Kelly Johnson's arrival at Lockheed in 1933 was soon followed by the start of the Second World War. Partly as a result of Johnson's work, World War II was the first major air war: airplanes killed 2.2 million people—more than 90 percent of them, around 2 million people, civilians, mainly women and children. The weapons used to defend against air attacks were crude and ineffective: antiaircraft guns that, on average, fired three thousand shells for each bomber they destroyed. As a result, nearly all bombers reached their targets. In the age of the nuclear bomb, this was a frightening statistic.

Immediately after the war, the new, urgent problem was how to defend against death from the sky, and the solution was surface-to-air missiles, which used the new technologies of computing and radar to locate, pursue, and destroy attacking aircraft. In Vietnam, the next major air war after World War II, surface-to-air missiles destroyed 205 U.S. planes, one for every 28 missiles fired—performance more than ten times better than the antiaircraft guns of World War II. Flying over enemy territory had become so dangerous it was almost suicidal.

This was the context for Ben Rich's "Hopeless Diamond" proposal. The paradigm for understanding aircraft had shifted. The problem now was not how to fly, or how to fly faster, but how to fly in secret.

The Hopeless Diamond was an attempt to solve that problem.

After bursting into Ben Rich's office, kicking him, and throwing his proposal on the floor, Kelly Johnson yelled, "This crap will never get off the ground."

Not every great innovator is a great manager of innovation. Johnson's yelling and screaming might have been the end of Rich's proposal but for one thing: the "Show me" rule.

The Skunk Works engineers had developed a tradition: when there was a dispute about something technical, they bet each other a quarter, then ran an experiment. Johnson and Rich had placed about forty of these bets during their years of working together. Johnson had won them all. There were two things Johnson always seemed to win: arm-wrestling matches—he'd worked as a brick carrier when he was young and had developed arms like thick ropes—and twenty-five-cent technical bets.

Rich said, "Kelly, this diamond is somewhere between ten thousand and one hundred thousand times lower in radar cross section than any U.S. military airplane or any new Russian MiG."

Johnson considered that. Lockheed had some experience building aircraft that evaded radar. In the 1960s, the company had developed an unmanned drone, called the D-21, that took photographs, dropped its camera to be picked up later, then blew itself up. The technology worked, but the program was a commercial failure. Johnson thought about the twelve-year-old drone that had failed, and said, "Ben, I'll bet you a quarter that our old D-21 drone has a lower cross section than that goddam diamond."

Or: "Show me." And so on September 14, 1975, the two men met in creation's equivalent of a duel.

Rich's team put a scale model of the Hopeless Diamond into an electromagnetic chamber and measured how hard it was to detect on radar—a quality the Lockheed engineers had started to call "stealth."

Rich and Johnson received the results and looked at them eagerly. The Hopeless Diamond was a thousand times stealthier than the D-21. Rich had won his first ever bet with Johnson. Johnson flipped Rich a quarter. Then he said, "Don't spend it until you see the damned thing fly."

The plane, code-named "Have Blue," did fly. It was the first-ever stealth aircraft, the parent of every subsequent undetectable aircraft, from the F-117 Nighthawk, to the MH-60 Black Hawk helicopters used in the 2011 raid on Osama bin Laden's compound in Pakistan, to the Lockheed SR-72, an almost invisible plane that flies at over forty-five hundred miles per hour. It was the product of an organization that valued action over talk, spent little time planning and lots of time trying, and resolved disputes about ideas not with arm wrestling or rank pulling but with two simple words: "Show me."

CHAPTER 9

GOOD-BYE, GENIUS

1 | THE INVENTION OF GENIUS

There is a desert more than a thousand miles long on Africa's Atlantic coast. Much of it is a sand sea, or *erg*, where the wind makes dunes twenty miles long and a thousand feet high. The desert is called the Namib, and it is home to a people called the Himba, whose women cover their skin and hair with milk fat, ash, and ocher, both for beauty and to protect themselves from the sun. In 1850, the Himba saw something odd in the dunes: men with white skin, covered in clothing, coming toward them through the sand. One of the men was thin and nervous. In time they found that he had a fetish for counting and measuring, and whenever he removed his Quaker-style "wide-awake" hat, they saw he had combed his hair over a bald spot that rose on his head like the moon. His name was Francis Galton. These people who had learned to live well in one of the world's most desolate places did not impress him. He wrote later that they were "savages" who needed to be "managed," whose food and possessions could be "seized," and who could not "endure the steady labour that we Anglo-Saxons have been bred to support."

Galton was one of the first Europeans to visit the Namib. He took his prejudices about the Himba and other African people he met back with him to England. After his half cousin Charles Darwin published *The Origin of Species,* in 1859, Galton became obsessed with it and started a career measuring and classifying humanity to promote selective breeding, an idea he eventually called "eugenics."

Galton's book *Hereditary Genius,* published in 1869, proposed that human intelligence was inherited directly and diluted by "poor" breeding. He later came to doubt his use of the word "genius" in the title, although it is not entirely clear why:

> There was not the slightest intention on my part to use the word genius in any technical sense, but merely as expressing an ability that was exceptionally high, and at the same time inborn. A person who is a genius is defined as a man endowed with superior faculties. The reader will find a studious abstinence throughout the work from speaking of genius as a special quality. It is freely used as an equivalent for natural ability. There is no confusion of ideas in this respect in the book, but its title seems apt to mislead, and if it could be altered now, it should appear as *Hereditary Ability.*

Geniuses, then, were not a species apart but men (*always* men, of course) with "superior natural ability." Galton is not specific about what geniuses have a superior natural ability *to do,* but he is very clear that *men like him* are far more likely to be endowed with this superior ability, whatever it is, than anyone else: "The natural ability of which this book mainly treats is such as a modern European possesses in a much greater average share than men of the lower races."

Lastly, and most importantly, this ability, while natural and bestowed mainly on "modern Europeans," could be improved with selective breeding: "There is nothing either in the history of domestic animals or in that of evolution to make us doubt that a race of men may be formed who shall be as much superior mentally and morally to the modern European, as the modern European is to the lowest of the Negro races."

Or: we can breed better people in the same way we can breed bigger cows.

The comparison to cows is not trite. Just as cows are graded using a classification system (in Britain, for example, an "E3" carcass is "excellent" and neither too lean nor too fat, while a "–P1" carcass is "poor" and skinny), so Galton proposed a grading system, or "Classification of Men According to Their Natural Gifts," ranging from "a," meaning something like "of below-average quality," to "X," for a one-in-a-million genius. The system, which Galton believed was "no uncertain hypothesis" but "an absolute fact," enabled him to make what he clearly thought were absolute comparisons between "races":

> The negro race has occasionally, but very rarely, produced such men as Toussaint l'Ouverture [the leader of the Haitian revolution of 1791], who are of our class F; that is to say, its X, or its total classes above G, appear to correspond with our F, showing a difference of not less than two grades between the black and white races, and it may be more. In short, classes E and F of the negro may roughly be considered as the equivalent of our C and D—a result which again points to the conclusion, that the average intellectual standard of the negro race is some two grades below our own.

This passage of Galton's makes no sense, and it is representative of his whole book. Without providing any evidence at all, he asserts that the best a black man can be is a "class F" type of cow, whereas the best a white man can be is a "class X" cow; this is two grades higher, and therefore white people are two grades better than black people. The best case against Galton's argument that white men are smarter than everyone else may be Galton's own stupidity.

But Galton was taken seriously. He gave centuries of prejudice a facade of reason and science. His work cast a dreadful shadow across the twentieth century, and into today. Galton's use of the word "genius" gave it the meaning it has now. To us, genius is what Galton said it was: a rare ability gifted by nature to a special few. You were either born with it or, more probably, you were not. But this was at best a secondary definition of genius in Galton's time. It was only because of the rise of eugenics, driven highest by the Nazis' belief in "racial hygiene," that the idea of genius as inherited superiority became common during the end of the nineteenth century and was the only accepted use by the end of the twentieth. There is a straight line from Galton's use of "genius" to Hitler's use of genocide.

A hypothesis is not false because it is offensive or atrocious. Galton's definition of genius as natural exceptional ability, reserved almost exclusively for white men, who then must ensure that only they father children for the good of the species, is not wrong because it is immoral; it is wrong because there is no evidence to support it. Galton's only evidence is self-evidence. His life's work was an elaboration of his prejudices, which were founded, as prejudices always seem to be, upon his conviction that he himself was one of a special breed.

All the evidence supports the opposite case: that natural ability is distributed among people of all types and is not the biggest factor determining our success. From the world-changing work of Edmond Albius, to the world-*saving* work of Kelly Johnson, we see that people everywhere can make differences big and small and that there is no way to guess who they will be in advance. When Rosalind Franklin revealed the human blueprint in DNA, she proved that there was nowhere for Galton's hypothetical racially determined exceptional ability to hide.

Genius as Galton defined it has no place in the twenty-first century—not because genius is not necessary but because we know it does not exist.

2 | ORIGINAL GENIUS

Long before Galton and eugenics, everyone had genius. The first definition of "genius" comes from ancient Rome, where the word meant "spirit" or "soul." This is the true definition of creative genius. Creating is to humans as flying is to birds. It is our nature, our spirit. Our purpose as a people and as individuals is to leave a legacy of new and improved art, science, and technology for future generations, just as our two thousand generations of ancestors did before us.

We are each a piece of something connected and complicated, something with such constant presence that it is invisible: the network of love and imagination that is the true fabric of humanity. This is not a fashionable view among people who claim to think. There is a false intellectual tradition of complaint that paints wonder as blunder, mistakes snorts for thoughts, and points at human beings as if they were mainly shameful. "But famine," "but war," "but Hitler," "but climate change": it is easier to look for flies in the soup than to work in the kitchen. But we *are* all connected, and we *are* creative. No one does anything alone. Even the greatest inventors build on the work of thousands. Creation is contribution.

We cannot know the weight of our contribution in advance. We must create for creation's sake, trust that our creations may have impacts we cannot foresee, and know that often the greatest contributions are the ones with the most unimaginable consequences.

The biggest consequence of our creation is us. The human population doubled between 1970 and 2010. In 1970, the average person lived to be fifty-two years old. In 2010, the average person lived to be seventy. Not only are twice as many people each living one-third longer, each individual's consumption of natural resources is increasing. Food intake was eight hundred thousand calories per person per year in 1970 and over a million calories a year in 2010. The amount of water we each consume more than doubled, from 160,000 gallons a year in 1970 to nearly 330,000 gallons a year in 2010. Despite the rise of the Internet and computers and the decline of printed newspapers and books, our use of paper increased from 55 pounds per person per year in 1970 to 120 pounds in 2010. We have more energy-efficient technology than we did in 1970, but we also have *more* technology, and more of the world has access to electricity, so, while we used 1,200 kilowatt-hours per person per year in 1970, we used 2,900 kilowatt-hours in 2010.

These changes are good for individuals right now: they mean more of us are living longer, healthier lives, with enough to eat and drink and a much better chance of avoiding or surviving illness and injury. The same is likely to be true of our children. But increased consumption will be a crisis for our entire species in the near future. It is not only how many of us there are and how much we each consume that is growing; the *rate of growth* of these numbers is growing, too. We are going faster *and* we are still accelerating. Our natural resources cannot grow as fast as our needs. If nothing changes, our species will one day ask for more than our planet can give; the only unknown is when.

These are not new concerns. In 1798, a book called *An Essay on the Principle of Population* was published in Britain. Its pseudonymous author warned of possible disaster:

The power of population is indefinitely greater than the power in the earth to produce subsistence for man. Population, when unchecked, increases in a geometrical ratio. Subsistence increases only in an arithmetical ratio. A slight acquaintance with numbers will show the immensity of the first power in comparison of the second. By that law of our nature which makes food necessary to the life of man, the effects of these two unequal powers must be kept equal. This implies a strong and constantly operating check on population from the difficulty of subsistence. This difficulty must fall some where, and must necessarily be severely felt by a large portion of mankind.

Or: we are producing more people than food, so the majority will soon starve.

The author was Thomas Malthus, a country vicar from the village of Wotton, thirty miles south of London. Malthus's father, inspired by French philosopher Jean-Jacques Rousseau, thought humanity was progressing toward perfection because of science and technology. Malthus the younger disagreed. His essay was a bleak picture painted to prove his father wrong.

Malthus was widely read and remained influential long after his death. Darwin and Keynes mentioned him favorably, Engels and Marx attacked him, and Dickens ridiculed him in *A Christmas Carol,* when Ebenezer Scrooge tells two gentlemen why he does not donate to the poor: "If they would rather die," said Scrooge, "they had better do it, and decrease the surplus population."

Or as Malthus put it:

The power of population is so superior to the power in the earth to produce subsistence for man, that premature death must in some shape or other visit the human race. The vices of mankind are active and able ministers of depopulation. But should they fail in this war of extermination, sickly seasons, epidemics, pestilence, and plague, advance in terrific array, and sweep off their thousands and ten thousands. Should success be still incomplete, gigantic inevitable famine stalks in the rear, and with one mighty blow, levels the population with the food of the world.

And yet we are still here.

Malthus was right about the growth in population—in fact, he greatly underestimated it. But he was wrong about its consequences.

At the close of the eighteenth century, when Malthus wrote his essay, there were almost a billion people in the world, the population having doubled in three centuries. This would have seemed like an alarming rate of growth to him. But in the twentieth century, the world population doubled twice more, reaching *two* billion in 1925 and *four* billion in 1975. According to Malthus's theory, this should have resulted in great famine. In fact, famine declined as population increased. In the twentieth century, 70 million people died due to lack of food, but most of them perished in the first few decades. Between 1950 and 2000, famine was eradicated from everywhere but Africa; since the 1970s, it has been concentrated in two countries: Sudan and Ethiopia. Fewer people are dying of starvation even though there are far more people on the planet.

The only way this famine could have been avoided, according to Malthus, was if "the vices of mankind" killed enough people first. In the first half of the twentieth century, that may have looked right: the First and Second World Wars combined to make the deadliest decades since the Black Death, killing as many as one in every four hundred people per year in the 1940s. But after that, war deaths dropped. From 1400 to 1900, about one in ten thousand people died in war every year, with peaks around 1600 and 1800, during the Wars of Religion and the

Napoleonic Wars. After 1950, that number is close to zero. Contrary to all Malthusian expectations, premature deaths plummet when population soars.

The reason is creation—or, more specifically, creators. When population grows, our ability to create grows even faster. There are more people creating, so there are more people with whom to connect. There are more people creating, so there are more tools in the tool chain. There are more people creating, so we have more time, space, health, education, and information for creating. Population is production. This is why there has been an apparent acceleration of innovation in the last few decades. We have not become innately more creative. There are just more of us.

And this is why we need new. Consumption is a crisis because of math; it is not yet a catastrophe because of creation. We beat change *with* change.

The chain of creation is many links long, and every link—each one a person creating—is essential. All stories of creators tell the same truths: that creating is extraordinary but creators are human; that everything right with us can fix anything wrong with us; and that progress is not an inevitable consequence but an individual choice. Necessity is not the mother of invention. You are.

ACKNOWLEDGMENTS

I owe a great debt to Robert W. Weisberg for his books *Creativity: Understanding Innovation in Problem Solving, Science, Invention, and the Arts* (2006), *Creativity: Beyond the Myth of Genius* (1993), and *Creativity: Genius and Other Myths* (1986); to Google; to Wikipedians everywhere; to the Internet Archive; to Christian Grunenberg, Alan Edwards, and Nathan Douglas for their artificial intelligence–driven database, DevonThink; and to Keith Blount and Ioa Petra'ka of Literature and Latte for Scrivener, their software for authors.

Most of the details in the story of Edmond Albius in chapter 1 come from Tim Ecott's book *Vanilla: Travels in Search of the Ice Cream Orchid.* Ecott did important primary research on Réunion to discover the true story of Edmond Albius.

Chapter 2 draws heavily from Lynne Lees's translation of *On Problem Solving,* by Karl Duncker. The description of a talk by Steve Jobs comes from "The 'Lost' Steve Jobs Speech from 1983," by Marcel Brown, published on Brown's *Life, Liberty and Technology* blog; a cassette tape from John Celuch; a transcript by Andy Fastow; and photographs by Arthur Boden provided by Ivan Boden.

Chapter 3's material about Judah Folkman is mainly from Robert Cooke's 2001 biography *Dr. Folkman's War* and a PBS documentary called *Cancer Warrior.* Stephen King's memoir *On Writing,* and an article titled "A Better Mousetrap," by Jack Hope, published in *American Heritage* magazine in 1996, were also essential sources for this chapter. The book referred to in the section "Strangers with Candy" is *Creativity: Flow and the Psychology of Discovery and Invention,* by Mihaly Csikszentmihalyi.

Robin Warren's 2005 Nobel Lecture, "Helicobacter: The Ease and Difficulty of a New Discovery," inspired chapter 4. Jeremy Wolfe of Brigham and Women's Hospital in Boston gave me an uncorrected prepublication proof of a paper he coauthored, "The Invisible Gorilla Strikes Again: Sustained Inattentional Blindness in Expert Observers," as well as other guidance on the subject of inattentional blindness. Robert Burton's book *On Being Certain* led me to many sources, including "Phantom Flashbulbs: False Recollections of Hearing the News About *Challenger,*" a 1992 paper by Ulric Neisser and Nicole Harsch. The full story of Dorothy Martin is in Festinger, Schachter, and Riecken's book *When Prophecy Fails.* There is more technical detail in Festinger's book *A Theory of Cognitive Dissonance.*

From chapter 5: Brenda Maddox's *Rosalind Franklin: The Dark Lady of DNA* is a wonderful biography of Franklin; Robert Merton's *On the Shoulders of Giants* is insightful and funny.

Chapter 6's description of the battle at William Cartwright's mill is informed by the Luddite Bicentenary blog, at ludditebicentenary .blogspot.co.uk; David Griffiths of the Huddersfield Local History Society in England helped with everything about the Luddites, especially by sending me Alan Brooke's and the late Lesley Kipling's book *Liberty or Death,* as well as many pamphlets. *The Amish,* by Donald Kraybill, Karen Johnson-Weiner, and Steven Nolt, was an invaluable source.

All work on motivation and creation by Teresa Amabile of Harvard

Business School, discussed in chapter 7, is wonderful, especially her 1996 book, *Creativity in Context*. The descriptions of and quotations from Woody Allen come mainly from Robert Weide's *Woody Allen: A Documentary* and Eric Lax's biography *Conversations with Woody Allen*. Annie Miler's café is Clementine, at 1751 Ensley Avenue in Los Angeles, California. I recommend the grilled cheese. Good luck parking.

There are many books about Lockheed's Skunk Works, which is described in chapter 8. Kelly Johnson's autobiography, *Kelly: More Than My Share of It All,* and Ben Rich's *Skunk Works: A Personal Memoir of My Years at Lockheed* both benefit from being by primary sources. Brian Jones's biography of Jim Henson and Michael Davis's *Street Gang* are excellent books about Henson and Oz. Lev Vygotsky's *Mind in Society* is still fascinating today. Tom Wujec maintains a website about the marshmallow challenge at marshmallowchallenge.com; I first heard about the challenge from my wonderful friend Diane Levitt, who learned about it from our mutual colleague Nate Kraft.

The data about famine in chapter 9 comes from *Famine in the Twentieth Century,* by Stephen Devereux, and the data about war comes from *The Better Angels of Our Nature,* by Steven Pinker. "Everything right with us can fix anything wrong with us" is a paraphrase from Bill Clinton's 1993 inaugural address, which was written mostly by Michael Waldman.

Early drafts of the sections "Obvious Facts," "Humanity's Choir," "A Can of Worms," and "Strangers with Candy" appeared in *Medium.*

Other references and sources are listed in the notes and bibliography, below. For additional information, see www.howtoflyahorse.com, which is an interactive companion for this book.

Many of the quotations in this book have been modified, with no change in meaning, to fit the text without the distraction of ellipses and square brackets; wherever possible, complete versions of these quotations are in the notes below. Some descriptive details in the text, such as facial expressions, are imagined or assumed; most, such as weather, are not. Most links in the notes use the URL-shortening service Bitly—indicated by the domain name "bit.ly"—and are simplified so that they can be typed into a web browser easily. The links will expand once entered and take you to the appropriate host site.

THANKS

Jason Arthur

Arlo Ashton

Sasha Ashton

Theo Ashton

Sydney Ashton

Elle B. Bach

Emma Banton

Julie Barer

Emily Barr

Larry Begley

Lizz Blaise

Aaron Blank

Lyndsey Blessing

Kristin Brief

Dick Cantwell

Katell Carruth

Amanda Carter

Henry Chen

Mark Ciccone

Paolo De Cesare

John Diermanjian

Larry Downes

Benjamin Dreyer

Mike Duke

Esther Dyson

Pete Fij

Stona Fitch

John Fontana

Andrew Garden

Audrey Gato

Tal Goretsky

Sarah Greene

Esther Ha

Alan Haberman

Mich Hansen

Adam Hayes

Nick Hayes

Chloe Healy

Rebecca Ikin

Durk Jager

Anita James

Gemma Jones

Levi Jones

Al Jourgensen

Mitra Kalita

Steve King

Pei Loi Koay

AJ Lafley

Cecilia Lee

Kate Lee

Bill Leigh

Diane Levitt

Maddy Levitt

Roxy Levitt

Gideon Lichfield

Angelina Fae Lukacin

John Maeder

Doireann Maguire

Yael Maguire

Sarah Mannheimer

Sylvia Massy
Sanaz Memarzadeh
Bob Metcalfe
Dan Meyer
Lisa Montebello
Alyssa Mozdzen
Jason Munn
Jun Murai
Eric Myers
Wesley Neff
Nicholas Negroponte
Christoph Niemann
Karen O'Donnell
Maureen Ogle
Ben Oliver
Sasha Orr
Sun Young Park
Shwetak Patel
Arno Penzias
John Pepper
Andrea Perry
Elizabeth Perry
Nancy Pine
Richard Pine
John Pitts
Elizabeth Price
Jamie Price
Kris Puopolo

Sin Quirin
David Rapkin
Nora Reichard
Matt Reynolds
Laura Rigby
Rhonda Rigby
Mark Roberti
Aaron Rossi
Kyle Roth
Eliza Rothstein
Paige Russell
Paul Saffo
Sanjay Sarma
Carsten Schack
Richard Schultz
Toni Scott
Arshia Shirzadi
Elizabeth Shreve
Tim Smucker
Bill Thomas
Bonnie Thompson
Adrian Tuck
Joe Volman
Pete Weiss
Marie Wells
Daniel Wenger
Ev Williams
Yukiko Yumoto

NOTES

PREFACE: THE MYTH

xiii   **In 1815, Germany's *General Music Journal* published a letter:** The letter was published in *Allgemeine Musikalische Zeitung,* or "General Music Journal," in 1815, vol. 17, pp. 561–66. For full descriptions of the Mozart letter hoax and its consequences, see Cornell University Library, 2002; Zaslaw, 1994; and Zaslaw, 1997.

xiv   **Mozart's real letters:** Mozart's compositional process is described by Konrad in Eisen, 2007; by Zaslaw in Morris, 1994; and in Jahn, 2013.

 xv   **In 1926, Alfred North Whitehead made a noun:** Many scholars have concluded that Whitehead invented the word "creativity" in Whitehead, 1926, within the following sentence: "The reason for the temporal character of the actual world can now be given by reference to the creativity and the creatures." Meyer, 2005, contains an excellent summary of this scholarship.

CHAPTER I: CREATING IS ORDINARY

 1   **A bronze statue stands in Sainte-Suzanne:** There is a picture of the statue of Edmond Albius at http://bit.ly/albiusstatue.

 1   **On Mexico's Gulf Coast, the people of Papantla:** The descriptions of vanilla and the story of Edmond Albius are based on Ecott, 2005, also Cameron, 2011.

 7   **The modern U.S. Patent and Trademark Office:** The first patent issued by what was then called the U.S. Patent Office was granted to Samuel Hopkins, an

inventor living in Pittsford, Vermont, for an improved way of making potassium
carbonate—in those days called "potash"—out of trees, mainly for use in soap,
glass, baking, and gunpowder. See Henry M. Paynter "The First Patent" (revised
version), http://bit.ly/firstpatent. The eight millionth patent was granted to
Robert Greenberg, Kelly McClure, and Arup Roy of Los Angeles for a prosthetic
eye that electrically stimulates a blind person's retina. See "Millions of Patents,"
USPTO, http://bit.ly/patentmillion. Actually, this was probably closer to the
8,000,500th patent issued, as the Patent Office started numbering patents in
series only in 1836.

7   **economist Manuel Trajtenberg:** See "The Mobility of Inventors and the
Productivity of Research," a presentation by Manuel Trajtenberg, Tel Aviv University, July 2006: http://bit.ly/patentdata. Using a multistage analysis of inventors' names, addresses, coinventors, and citations, Trajtenberg ascertained that
the 2,139,313 U.S. patents granted at the time of his analysis had been issued to
1,565,780 distinct inventors. The granted patents had a mean of 2.01 inventors
per patent. Trajtenberg's analysis suggests that the average number of patents per
inventor is 2.7. By taking the 2011 number of 8,069,662, multiplying it by 2.01 to
get the total named inventors, and then dividing by 2.7 to account for the average
number of patents per inventor, I calculated that there were around 6,007,415
unique inventors named on granted patents by the end of 2011.

7   **The inventors are not distributed evenly:** This analysis assumes that Trajtenberg's numbers are constants, and so the number of "inventors" scales in exactly the
same way as the number of patents, as published by the USPTO and cited above.

7   **Even with foreign inventors removed:** My own analysis, using USPTO data
as cited above, U.S. Census data, and Trajtenberg's numbers as constants. The
USPTO started tracking patents awarded to foreign residents in 1837. The figure
of 1,800 is six in a million, so closer to 1 in 166,666, but I rounded up to a clean
number to keep both statistics as "1 in" something.

8   **In 1870, 5,600 works were registered for copyright:** See the Annual Report of
the Librarian of Congress, 1886: http://bit.ly/copyrights1866.

8   **In 1946, register of copyrights Sam Bass Warner:** See the 49th Annual Report
of the Register of Copyrights, June 30, 1946: http://bit.ly/copyrights1946.

8   **In 1870, there was 1 copyright registration for every 7,000:** History of registrations taken from Annual Report of the Register of Copyrights, September 30,
2009: http://bit.ly/copyrights2009. The analysis is my own, using U.S. Census
data. In 1870, there were 3 registrations for every in 20,000 people, which I
rounded to 1 registration for every 7,000 people to match the format of following
number, 1 in 400.

8   **one for every 250 U.S. citizens:** Data about the *Science Citation Index* from
Eugene Garfield, "Charting the Growth of Science," paper presented at the
Chemical Heritage Foundation, May 17, 2007; http://bit.ly/garfieldeugene. The
analysis is my own, using U.S. Census data.

9   **a typical NASCAR race:** The average NASCAR race attendance in 2011 was
98,818, based on data from ESPN / Jayksi LLC, at http://bit.ly/nascardata. The
number of U.S. residents granted first patents in 2011 was 79,805, based on
USPTO data and Trajtenberg's constants.

10 **five African wildcats:** See Driscoll et al., 2007.

11 **Dolphins use sponges to hunt for fish:** Krützen et al., 2005.

11 **human tools were monotonous for a million years:** Mithen, 1996, and Kuhn and Stiner in Mithen, 2014.

13 **"Despite great qualitative and quantitative":** Casseli, 2009.

15 **"The most fundamental facts":** Ashby, 1952.

15 **A San Franciscan named Allen Newell:** See Newell, "Desires and Diversions," a lecture presented at Carnegie Mellon, December 4, 1991; the video is available at http://bit.ly/newelldesires, courtesy of Scott Armstrong.

16 **"The data currently available about the processes":** Newell, 1959. Available at http://bit.ly/newellprocesses.

16 **Weisberg was an undergraduate during the first years:** Robert Weisberg's résumé is available at http://bit.ly/weisbergresume.

17 **"when one says of someone that":** Weisberg, 2006.

17 **Sprengel's peers did not want to hear that flowers had a sex life:** Zepernick and Meretz, 2001.

18 **Titles available in today's bookstores:** Titles found at Amazon.com.

18 **Weisberg's books are out of print:** According to Amazon.com, where only a Kindle edition of Weisberg's last, more academic title, *Creativity: Understanding Innovation in Problem Solving, Science, Invention, and the Arts,* is available "new."

18 **"creativity now is as important in education as literacy":** Ken Robinson, TED talk, June 27, 2006. Transcript at http://bit.ly/robinsonken.

18 **Cartoonist Hugh MacLeod:** MacLeod, 2009.

19 **The best-known version was started in 1921:** See Terman's own work, especially Terman and Oden, 1959. Shurkin, 1992, offers an excellent review of Terman's work.

22 **"within the reach of everyday people in everyday life":** Torrance, 1974, quoted in Cramond, 1994.

24 **open our veins and bleed:** Versions of this comment have been attributed to several writers. According to Garson O'Toole, the original is "It is only when you open your veins and bleed onto the page a little that you establish contact with your reader," from "Confessions of a Story Writer," by Paul Gallico, 1946; http://bit.ly/openavein.

25 **bestselling literary series was begun by a single mother:** J. K. Rowling; see http://bit.ly/rowlingbio.

25 **a career more than fifty novels long:** "Four years before, I had been running sheets in an industrial laundry for $ 1.60 an hour and writing Carrie in the furnace-room of a trailer," King, 2010. See also Lawson, 1979.

25 **world-changing philosophy was composed in a Parisian jail:** Paine, 1794.

25 The **"man with a permanent position as a patent examiner"** was Albert Einstein.

CHAPTER 2: THINKING IS LIKE WALKING

26 **Thomas Mann prophesied the perils of National Socialism:** Mann, 1930.

27 **He made two applications to become a professor:** Now the Humboldt University of Berlin (German: Humboldt-Universität zu Berlin), founded in 1810 as the University of Berlin (Universität zu Berlin). In Duncker's day it was known

as the Frederick William University (Friedrich-Wilhelms-Universität), and later (unofficially) also as the Universität Unter den Linden.

27 **Both were rejected:** This and other Duncker biographical details from Schnall, 1999, published in Valsiner, 2007; see also Simon, 1999, in Valsiner, 2007.

27 **He published his masterwork, *On Problem Solving*:** Duncker, 1935. Translation: Duncker and Lees, 1945.

27 **"Today the sun is brilliantly shining":** Quotation edited for length and clarity from Isherwood, 1939. The unedited passage is: "To-day the sun is brilliantly shining; it is quite mild and warm. I go out for my last morning walk, without an overcoat or hat. The sun shines, and Hitler is master of this city. The sun shines, and dozens of my friends—my pupils at the Workers' School, the men and women I met at the I.A.H.—are in prison, possibly dead. But it isn't of them that I am thinking—the clear-headed ones, the purposeful, the heroic; they recognized and accepted the risks. I am thinking of poor Rudi, in his absurd Russian blouse. Rudi's make-believe, story-book game has become earnest; the Nazis will play it with him. The Nazis won't laugh at him; they'll take him on trust for what he pretended to be. Perhaps at this very moment Rudi is being tortured to death. I catch sight of my face in the mirror of a shop, and am shocked to see that I am smiling. You can't help smiling, in such beautiful weather. The trams are going up and down the *Kleistsrrasse,* just as usual. They, and the people on the pavement, and the teacosy dome of the *Nollendortplatz* station have an air of curious familiarity, of striking resemblance to something one remembers as normal and pleasant in the past—like a very good photograph."

28 **an immigrant who'd left the tiny Lithuanian village of Sventijánskas:** Detail from Kimble, 1998, in which *Sventijánskas* is transliterated as "Swiencianke." Krechevsky was born Yitzhok-Eizik Krechevsky and started using the first name Isadore when he attended school in the United States.

28 **The joint paper, "On Solution-Achievement":** Duncker, 1939.

29 **Duncker published his second paper, on the relationship between familiarity and perception:** Duncker, 1939b.

29 **Duncker's third paper of the year:** Duncker, 1939c.

29 **He drove to nearby Fullerton:** *New York Times,* 1940.

30 **In Berkeley, the University of California awarded:** Rensberger, 1977.

32 **"If a situation is introduced in a certain perceptual structure":** This is my translation—the Lees translation uses "structuration" instead of "structure."

32 **Psychologists and people who write about creation:** Duncker's *On Problem Solving* has around twenty-two hundred citations, according to Google scholar: http://bit.ly/dunckercitations.

33 **How did Charlie die?:** Weisberg, 1986.

33 **the Prisoner and Rope Problem:** Described in Metcalfe, 1987, cited in Chrysikou, 2006, and Weisberg, 2006.

34 **This is the source of the cliché "thinking outside of the box":** See http://bit.ly /outsideofbox. A possible alternative origin story involves a man smuggling bicycles by distracting border guards with a box of sand balanced on the handlebars.

34 **It is a summary of a Sherlock Holmes story:** "The Adventure of the Speckled Band" in Doyle, 2011.

35 **The surprising solution that a snake killed her follows:** Doyle may have

made a mistake in this story. When Doyle wrote "The Adventure of the Speckled Band," in 1892 it was generally believed that snakes were deaf. This led to much speculation among Holmes's enthusiasts about what kind of snake Doyle had in mind, or whether it was in fact, a lizard. Later research, starting in 1923, and culminating as recently as 2008, showed that snakes can hear, via their jaws, despite not having external ears.

35 **Many people do not think using words:** See Weisberg, 1986, and Chrysikou, 2006, for examples of how this has been established.

35 **Robert Weisberg asked people to think aloud:** Weisberg and Suls, 1973.

37 **Six undergraduates talking their way through a puzzle:** Weisberg and Suls, 1973. Weisberg's paper describes six related experiments, one of which was evaluating solutions to the problem rather than solving it; 376 is the number of subjects who participated in the other five experiments.

38 **Oprah Winfrey has trademarked it:** Winfrey's Harpo companies own two "live" trademarks using the phrase "aha! moment," registration number 3805726 and registration number 3728350.

38 **Greek general Hiero was crowned king of Syracuse:** Vitruvius, 1960.

38 **Hiero asked Syracuse's greatest thinker:** Biello, 2006.

39 **Galileo pointed this out in a paper called "La Bilancetta":** Galileo, 2011. Translation from Fermi and Bernardini, 2003.

39 **Buoyancy, not displacement:** This is explained beautifully by Chris Rorres at http://bit.ly/rorres.

39 **But let's take Vitruvius's story at face value:** Vitruvius, 1960.

40 **"In the summer of the year":** Edited from Coleridge, 2011. The complete quotation is:

> In the summer of the year 1797, the Author, then in ill health, had retired to a lonely farm-house between Porlock and Linton, on the Exmoor confines of Somerset and Devonshire. In consequence of a slight indisposition, an anodyne had been prescribed, from the effects of which he fell asleep in his chair at the moment that he was reading the following sentence, or words of the same substance, in "Purchas's Pilgrimage": "Here the Khan Kubla commanded a palace to be built, and a stately garden thereunto. And thus ten miles of fertile ground were inclosed with a wall." The Author continued for about three hours in a profound sleep, at least o the external senses, during which time he has the most vivid confidence, that he could not have composed less than from two to three hundred lines; if that indeed can be called composition in which all the images rose up before him as things, with a parallel production of the correspondent expressions, without any sensation or consciousness of effort. On awaking he appeared to himself to have a distinct recollection of the whole, and taking his pen, ink, and paper, instantly and eagerly wrote down the lines that are here preserved. At this moment he was unfortunately called out by a person on business from Porlock, and detained by him above an hour, and on his return to his room, found, to his no small surprise and mortification, that though he still retained some vague and dim recollection of the general purport of the vision, yet, with the exception of some eight or ten scattered lines and images, all the rest had passed away like the images on the surface of a stream into which a stone has been cast, but, alas! without the after restoration of the latter!

40 **Coleridge used a similar device—a fake letter from a friend:** See Coleridge, 1907, where a letter from a "friend" interrupts chapter 13 of his *Biographia Literaria.* Bates, 2012, describes the "friend" as "a humorous gothic counterfeit."

41 **Coleridge says the poem was "composed in a sort of reverie":** Hill, 1984.

41 **"I was sitting writing at my textbook but the work":** Benfey, 1958.

42 **This is a case of visual imagination helping solve a problem:** Based on Weisberg, 1986 and Rothenberg, 1995.

42 **A sudden revelation has also been attributed to Einstein:** Einstein, 1982.

42 **In Einstein's own words: "I was led to it by steps":** Moszkowski, 1973, p. 96. The complete quotation is "But the suddenness with which you assume it to have occurred to me must be denied. Actually I was lead [*sic*] to it by *steps* arising from the *individual* laws."

42 **These psychologists conducted hundreds of experiments:** Hélie, 2012, includes references to many of these experiments. Advocates of the "incubation" hypothesis now use the term "implicit cognition."

42 **They showed the subject pictures of entertainers:** Read, 1982. "The research was initiated while both authors were on study leave at the University of Colorado." Cited and discussed in Weisberg, 1986.

43 **Other studies into the feeling:** e.g., Nisbett, 1977.

43 **In one experiment he sorted 160 people:** Olton and Johnson, 1976.

44 **He designed a different study:** Olton, 1979. Cited in Weisberg, 1993.

44 **Most researchers now regard incubation as folk psychology:** The phrase "folk psychology" is used in Vul, 2007. See also Dorfman et al., 1996; Weisberg, 2006, which contains a thorough review of studies of incubation; Dietrich and Kanso, 2010; and Weisberg, 2013, which critiques attempts to study insight using neural imaging and also analyzes the popularization of incubation by journalist Jonah Lehrer. Incubation is not completely discredited, however; some psychologists are reviving the hypothesis under the name "implicit cognition." Weisberg, 2014, attempts to incorporate theories of incubation into theories of ordinary thinking.

44 **"'Why doesn't it work?' or, 'What should I change to make it work?'":** Duncker, 1945. I have adjusted the translation—the Lees translation uses "alter" instead of "change."

45 **When Jobs announced Apple's first cell phone:** Talk by Steve Jobs at MacWorld San Francisco on January 9, 2007. Video: http://bit.ly/keyjobs. Transcript by Todd Bishop and Bernhard Kast: http://bit.ly/kastbernhard.

45 **Apple sold 4 million phones in 2007:** Data from Apple Inc. annual reports summarized at http://bit.ly/salesiphone. Adjusted to convert fiscal years to calendar years and rounded to the nearest million.

46 **This was true, irrelevant, and revealing:** The microphone built into the original iPhone had a narrow frequency response of about 50Hz to about 4kHz, compared, for example, to the subsequent iPhone 3G, which ranged from below 5Hz to 20kHz. Analysis by Benjamin Faber at http://bit.ly/micriphone.

46 **"If it ain't broke, don't fix it":** This phrase was popularized by Bert Lance, director of the Office of Management and Budget in the Carter administration, who used it in 1977. See *Nations Business,* May 1977, p. 27, at http://bit.ly/dontfix.

46 **Korean electronics giant LG launched:** The LG Prada, or LG KE850, was announced in December 2006 and made available for sale in May 2007. Apple

announced the iPhone in January 2007 and made it available for sale in June 2007. The LG Prada was the first cell phone with a capacitive touch screen. See http://bit.ly/ke850.

47 **The secret of Steve was evident in 1983:** This was the IDCA, or International Design Conference Aspen, 1983. The IDCA is now part of the Aspen Design Summit, organized by the American Institute of Graphic Arts. More at http://bit.ly/aspendesign.

47 **"If you look at computers, they look like garbage":** Brown, 2012, based on a cassette tape from John Celuch of Inland Design and a transcription by Andy Fastow at http://bit.ly/jobs1983. The transcription has been edited slightly for clarity. The description of Jobs's appearance is based on photos by Arthur Boden, posted by Ivan Boden at http://bit.ly/ivanboden.

48 **"One minute he'd be talking about sweeping ideas":** Mossberg, 2012.

48 **One symbol lived long after the cat:** TV Tropes, at http://bit.ly/felixbulb. Many Felix the Cat cartoons showing prop use are available online — see, for example, http://bit.ly/felixcartoon.

48 **Psychologists adopted the image:** Wallas, 1926.

49 **The most famous is brainstorming:** Osborn, 1942. See also http://bit.ly/alexosborn.

49 **"Brainstorming is often used in a business setting":** Extract from "Brain-storming Techniques: How to Get More Out of Brainstorming" at http://bit.ly/mindtoolsvideo. Transcript at http://bit.ly/manktelow.

50 **Researchers in Minnesota tested this:** Dunnette, 1963. Cited in Weisberg, 1986.

50 **Follow-up research tested whether larger groups:** Bouchard, 1970. Cited in Weisberg 1986.

50 **Researchers in Indiana tested this by asking groups of students:** Weisskopf-Joelson and Eliseo, 1961. Cited in Weisberg, 1986.

50 **Subsequent studies have reinforced this:** See, for example, Brilhart, 1964, as discussed by Weisberg, 1986.

50 **Steve Wozniak, Steve Jobs's cofounder at Apple:** Wozniak, 2007. Cited in Cain, 2012.

51 **According to novelist Stephen King:** King, 2001.

51 **political scientists William Ogburn and Dorothy Thomas:** Ogburn and Thomas, 1922.

52 **the moon chewed the sun in a partial solar eclipse:** Eclipse details at http://bit.ly/rhinoweclipse.

52 **"Sacrifices must be made":** Details of Lilienthal's death from Wikipedia, at http://bit.ly/lilienthalotto.

52 **"The balancing of a flyer may seem":** Wright, 2012.

53 **They saw an airplane as "a bicycle with wings":** Heppenheimer, 2003. Cited in Weisberg, 2006.

54 **"a sport to which we had devoted so much attention":** Wright, 2012.

54 **"Having set out"** through **"our own measurements":** Wright, 2012.

55 **Wings needed to be much bigger:** The Wrights' math was correct. Today, aero-dynamicists use a Smeaton coefficient of 0.00327. See Smithsonian National Air and Space Museum at http://bit.ly/smeatoncoeff.

55 **The Wrights' aircraft are the best evidence:** This point is beautifully illus-
trated in a presentation called "Invention of the Airplane" from NASA's Glenn
Research Center, available at http://bit.ly/manywings. See especially slide 56.

56 **On November 1, 1913, Franz Kluxen entered:** Little is known about Franz
Kluxen of Münster (also listed in catalogs as Kluxen of Boldixum, a district of
Wyk, a town on the German island of Föhr in the North Sea). According to Richardson, 1996, Kluxen may have been "one of the earliest (he started in 1910) and
most serious buyers of Picasso in pre-1914 Germany . . . By 1920, all the Kluxen
Picassos that can be traced had changed hands. Kluxen may have been a victim of
the war or of hard times."

56 **He forbade artificial chalk made from gypsum:** Natural chalk is from the Cre-
taceous period, circa 145.5 + 4 to 65.5 + 0.3 million years ago; it includes ancient
cell fragments visible only by microscope. Steele et al., in Smithgall, 2011.

56 **The picture covered thirty square feet:** *Painting with White Border* is 140 cm x
200 cm = 2.8 square meters = 30.14 square feet.

57 **"extremely powerful impressions I had experienced":** Kandinsky, *"Picture
with the White Edge,"* in Lindsay and Vergo, 1994. Cited in Smithgall, 2011.

57 **a common Kandinsky motif and a symbol:** See, for example, Kandinsky's
Painting with Troika, 1911. The troika symbolizes divinity by recalling the prophet
Elijah's fiery chariot ride to heaven.

59 **His first works, painted in 1904:** See, for example, *Russische Schöne in Landschaft,*
from around 1904.

59 **His last, painted in 1944:** See, for example, *Gedämpfter Elan,* 1944.

CHAPTER 3: EXPECT ADVERSITY

60 **One summer night in 1994, a five-year-old named Jennifer:** Jennifer is
real—I have omitted her last name to help protect her privacy—and so are all the
important details of her story. A few narrative details—that she had a pretty face,
that her father signed the consent form, that she cried when given her shots—are
imagined or assumed. The sources for Judah Folkman's story are Cooke, 2001;
Linde, 2001; published academic papers; and Folkman's obituaries.

62 **"I had seen and handled cancers":** Linde, 2001.

63 **Scientists had little respect for surgeons:** Today it is common for the best
medical doctors to also do basic research. Judah Folkman is one reason why. For
data on the rise of physician-scientists from the 1970s onward, see Zemlo, 2000.

65 **Angiogenesis became an important theory:** One promising line of investigation is whether regular doses of aspirin and other medicines modulate angiogenesis and reduce the risk of, for example, colon cancer, lung cancer, breast cancer,
and ovarian cancer. See Albini et al., 2012; Holmes et al., 2013; Tsoref et al., 2014;
and Trabert et al., 2014.

66 **A journey of a thousand miles ends with a single step:** The famous line "A
journey of a thousand miles starts with single step" is from chapter 64 of the *Tao
Te Ching;* Tsu, 1972.

66 **He knows how many e-mails he has sent:** Wolfram, 2012. It is a year and a half
or writing and deleting because 7 deletes out of a every 100 keystrokes is 7 per-

cent, but you also have 7 percent of keystrokes then getting deleted; this means 14 percent of keystrokes result in no extra text; 14 percent of ten years rounded up is a year and a half. This assumes that, on average, it takes as long to decide to delete something as it does to decide to write it.

66–67 **Stephen King, for example, has published:** Includes novels, screenplays, collections of short stories, and works of nonfiction. From Wikipedia's Stephen King bibliography at http://bit.ly/kingbibliography.

67 **He says he writes two thousand words a day:** King, 2001. "I like to get ten pages a day, which amounts to 2,000 words."

67 **Between the beginning of 1980 and the end of 1999:** My count of Stephen King's words starts with *Firestarter* (1980) and ends with *The New Lieutenant's Rap* (1999); it excludes the unedited version of *The Stand,* which is essentially a reprint of a prior book, and *Blood & Smoke,* which is King reading stories published elsewhere. I used the page count in the Wikipedia bibliography at http://bit.ly /kingbibliography, which is for the hardback format of each book, and I assumed three hundred words per page. I subtracted six months because King was injured and barely writing after June 1999. King did not begin writing his *Entertainment Weekly* column, "The Pop of King," until 2003, so that does not count toward his word total.

67 **"That DELETE key is on your machine":** King, 2001.

67 **One of King's most popular books:** King, 2001: "This is . . . the one my long-time readers still seem to like the best."

67 **"twelve hundred pages long and weighed twelve pounds":** King, 2010.

67 **"If I'd had two or even three hundred pages":** King, 2001.

67 **"There's a misconception that invention":** From Dyson's website, at http://bit .ly/dysonideas.

68 **"just an ordinary person":** Dyson interviewed at a WIRED Business Conference, 2012. Video at http://bit.ly/videodyson.

68 **"The north and south winds met":** Baum, 2008.

69 **house dust particles about a millionth of a meter wide:** House dust dimensions from "Diameter of a Speck of Dust" in *The Physics Factbook,* edited by Glenn Elert, written by his students at http://bit.ly/dustsize, with sense checking by Matt Reynolds of the University of Washington.

69 **"I'm a huge failure because I made 5,126 mistakes":** Dyson interviewed at a Wired Business Conference, 2012. Video at http://bit.ly/videodyson.

69 **"I wanted to give up almost every day":** Edited from Dyson's website, at http:// bit.ly/dysonstruggle. Full quotation: "I wanted to give up almost every day. But one of the things I did when I was young was long distance running, from a mile up to ten miles. They wouldn't let me run more than ten miles at school—in those days they thought you'd drop down dead or something. And I was quite good at it, not because I was physically good, but because I had more determination. I learned determination from it. A lot of people give up when the world seems to be against them, but that's the point when you should push a little harder. I use the analogy of running a race. It seems as though you can't carry on, but if you just get through the pain barrier, you'll see the end and be okay. Often, just around the corner is where the solution will happen."

69 **a personal fortune of more than $5 billion:** Dyson Ltd.'s 2013 revenues esti-

mated at £6 billion by Wikipedia, at http://bit.ly/dysoncompany. Dyson's net worth was estimated at £3 billion by the *Sunday Times* in 2013. See http://bit.ly/dysonworth.

69 **"Iterative Process":** Rubright, 2013.

69 **"Try again. Fail again. Fail better":** Beckett, 1983.

70 **A Hungarian psychology professor once wrote:** Csikszentmihalyi, 1996.

71 **"It is only half an hour":** Letter from Charles Dickens to Maria Winter, written on April 3, 1855, published in Dickens, 1894. Appears in Amabile, 1996, citing Allen, 1948. The complete quotation is: "I hold my inventive capacity on the stern condition that it must master my whole life, often have complete possession of me, make its own demands upon me, and sometimes, for months together, put everything else away from me. If I had not known long ago that my place could never be held, unless I were at any moment ready to devote myself to it entirely, I should have dropped out of it very soon. All this I can hardly expect you to understand—or the restlessness and waywardness of an author's mind. You have never seen it before you, or lived with it, or had occasion to think or care about it, and you cannot have the necessary consideration for it. 'It is only half an hour,'—'It is only an afternoon,'—'It is only an evening,' people say to me over and over again; but they don't know that it is impossible to command one's self sometimes to any stipulated and set disposal of five minutes,—or that the mere consciousness of an engagement will sometimes worry a whole day. These are the penalties paid for writing books. Whoever is devoted to an art must be content to deliver himself wholly up to it, and to find his recompense in it. I am grieved if you suspect me of not wanting to see you, but I can't help it; I must go my way whether or no."

73 **Semmelweis had convincing data to support his hypothesis:** Ignaz Semmelweis was not the only doctor to suspect that puerperal fever was being transmitted to patients by doctors. He didn't know it, but he was one of several physicians who had reached the same conclusion. Fifty years earlier, in Scotland, a surgeon named Alexander Gordon wrote about it; in 1842, Thomas Watson, a professor at the University of London, started recommending hand-washing; and in 1843, American Oliver Wendell Holmes published a paper about it. All were ignored or condemned.

73 **Semmelweis saved the lives:** Data from Semmelweis, 1859. Semmelweis's numbers are not entirely clear, and there is no way to know exactly how many women would have died without hand-washing. The mean patient death rate in the First Clinic in the fourteen years before hand-washing was introduced was 8 percent, versus 3 percent in the Second Clinic over the same period. The average death rate in the First Clinic dropped to 3 percent in the years 1846 (when hand-washing was introduced in May), 1847 and 1848 (when Semmelweis was terminated in March). If the average death rate in the First Clinic had remained at 8 percent in these three years, then 548 more women would have died. This is the basis for the statement that Semmelweis "saved the lives of around 500 women." This number is undoubtedly low. It does not include the fact that the average death rate only returned to its pre-hand-washing levels several years after Semmelweis's dismissal, nor, as mentioned, does it include babies, as there is not enough data about newborns in Semmelweis's paper to estimate how many

babies were saved. (I have assumed that Semmelweis's use of the term "patients" in his data about deaths means the numbers refer to women only, as this is how he uses the word elsewhere in the paper.)

75 **"A wise man proportions his belief to the evidence"**: The Hume quotation is from Hume, 1748; the Sagan quotation is from the opening lines of the PBS television program *Cosmos*, episode 12, first aired on December 14, 1980, available at http://bit.ly/extraordinaryclaims; the Truzzi quotation is from Truzzi, 1978. Laplace's quotation has a more complicated pedigree. The original source is Laplace, 1814, which states, "We are so far from knowing all the agents of nature and their diverse modes of action that it would not be philosophical to deny phenomena solely because they are inexplicable in the actual state of our knowledge. But we ought to examine them with an attention all the more scrupulous as it appears more difficult to admit them." This was rewritten as "The weight of the evidence should be proportioned to the strangeness of the facts" and called "The Principle of Laplace," by Théodore Flournoy in Flournoy, 1900, but is most commonly repeated as "The weight of evidence for an extraordinary claim must be proportioned to its strangeness." All four quotations are cited in the Wikipedia entry for Truzzi, at http://bit.ly/marcellotruzzi; the story of Laplace's quotation is told in the Wikipedia entry for Laplace, at http://bit.ly/laplacepierre.

76 **"If a man has good corn, or wood, or boards"**: Emerson, 1909.

76 **"If a man can write a better book"**: Yule, 1889. Cited in Hope, 1996.

76 **"Build a better mousetrap"**: Much of this section is based on a brilliant article by Jack Hope. Hope, 1996.

76 **More than five thousand mousetrap patents**: Hope, writing in 1996, estimated 4,400 patents, growing at 40 a year. His projection appears to be accurate: by May 2014 there were around 5,190 mousetrap patents, and applications show no sign of slowing down. See: http://bit.ly/mousetraps.

77 **Almost all of them cite the quotation**: Hope, 1996, quoting Joseph H. Bumsted, vice-president of mousetrap manufacturer Woodstream Corporation: "They feel it was written just for them, and they recite it as if that in itself were reason for Woodstream to buy their ideas!"

77 **Emerson could not have written it**: Emerson died in 1882; the first mousetrap patent was issue in 1894.

77 **This changed in the late 1880s**: Hooker's mousetrap has U.S. Patent Number 0528671. See http://bit.ly/hookertrap.

77 **Hooker's "snap trap" was perfected within a few years**: See http://bit.ly /victortrap. In May 2014, you could buy 20 traps for $15, with free shipping.

78 **"Nation's most precious natural resource"**: Ergenzinger, 2006.

78 **One company, Davison & Associates**: At first, Davison was ordered to pay $26 million in compensation. The FTC and the company then reached a settlement and Davison made a "non punitive" payment of $10.7 million, which I rounded up to $11 million to keep the prose simple.

79 **Many of their inventions are based on Davison's own ideas**: See, for example, the "Swingers Slotted Spoon" at http://bit.ly/davisonspoon. Despite being listed on the "Samples of Client Products" section of the Davison Web site the product information reveals: "This corporate product was invented and licensed by Davison for its own benefit."

79 **sales of $45 million a year:** This is calculated based on the disclosed number of 11,325 people a year buying a "pre-development agreement" at the published price of $795 and the disclosed number of 3,306 people buying a "new product sample agreement" at $11,500 which is halfway between the published estimated price of $8,000 – $15,000. This adds up to gross annual revenue from these services of $47,022,375. Sources: http://www.davison.com/legal/ads1.html, and http://www.davison.com/legal/aipa.html, viewed and saved on December 31, 2012. "Other public information" refers to Dolan, 2006: "Last year [presumably 2005], he [George Davison] says, his shop netted $2 million on $25 million in revenue." http://bit.ly/dolankerry.

80 **He denounced Hervieu's use of a dummy as a "sham":** The word in French is *chiqué*, which could also be translated as "bluff," or "deception." Reichelt: *"Je veux tenter l'expérience moi-même et sans chiqué [sic], car je tiens à bien prouver la valeur de mon invention."* *Le Petit Journal*, February 5, 1912, "L'Inventeur Reichelt S'est Tué Hier," at http://bit.ly/petitjournal.

80 **Reichelt had made sure his test:** He met with journalists the evening before the jump; the Pathé news footage of his jump, which was never aired, is at http://bit.ly/reicheltjump. The description of Reichelt's preparations, leap, and subsequent death, are based on this film.

80 **"I am so convinced my device will work properly":** Edited and translated from the French: *"Je suis tellement convaincu que mon appareil, que j'ai déjà experimenté, doit bien fonctionner, que demain matin, après avoir obtenu l'autorisation de la préfecture de police, je tenterai l'expérience du haut de la première plateforme de la Tour Eiffel."* From *Le Petit Journal*, February 5, 1912, "L'Inventeur Reichelt S'est Tué Hier," at http://bit.ly/petitjournal.

81 **Reichelt fell for four seconds:** Calculated from Green Harbor Publications, "Speed, Distance, and Time of Fall for an Average-Sized Adult in Stable Free Fall Position," 2010, at http://bit.ly/fallspeed.

82 **Hervieu was not the only one:** *Le Matin*, February 5, 1912 (number 10205), *"Expérience tragique,"* at http://bit.ly/lematin: *"La surface de votre appareil est trop faible, lui disait-on; vous vous romprez cou"*—"The surface of your device is too small, he was told; you will break your neck."

82 **"For a successful technology":** From Volume 2, Appendix F, of the United States Presidential Commission on the Space Shuttle Challenger Accident, 1986, at http://bit.ly/feynmanfooled.

83 **In the 1950s, two psychologists:** "High school" is assumed based on the grade level and birth years the children mentioned in their autobiographical essays. Getzels 1962. There were 533 children in total.

83 **Getzels and Jackson found that the most creative students:** These are all bright children to begin with. The mean IQ at the school was around 135. The difference in IQ scores between the "most creative" and "least creative" here is relative to their peers.

84 **It has been repeated many times:** See, for example, Bachtold, 1974; Cropley, 1992; and Dettmer, 1981.

84 **98 percent:** Feldhusen, 1975. Cited in in Westby, 1995. Westby also hypothesizes that teachers favor less creative childen over more creative children in part because more creative children tend to be harder to control.

84   **The Getzels-Jackson effect is not restricted:** Staw, 1995.

85   **In one experiment, Dutch psychologist Eric Rietzschel:** Rietzschel, 2010,
      Study 2.

85   **When Rietzschel asked people to assess their own work:** Rietzschel, 2010,
      Study 1.

86   **When we are in familiar situations:** Gonzales, 2004: "Normally, hippocampal
      cells fire perhaps only once every second on average. But at that mapped place,
      they fire hundreds of times faster."

86   **Uncertainty is an aversive state:** See, for example, Heider, 1958; Whitson, 2008

86   **Psychologists can show this in experiments:** See, for example, Mueller, 2012.

86   **rejection hurts:** For the neural basis of why this is so, see Eisenberger, 2004;
      Eisenberger, 2005.

86   **comes from the Old English *spurnen,* "to kick":** "Old English" means English
      spoken from the mid-fifth through mid-twelfth centuries.

86   **In 1958, psychologist Harry Harlow proved:** Artistotle, 2011, VIII.1155a5:
      "Without friends no one would wish to live, even if he possessed all other goods."
      Cited in Eisenberger and Lieberman, 2004.

87   **We know we should not suggest:** Flynn and Chatman, 2001; Runco, 2010. Both
      cited in Mueller et al., 2012.

87   **"Luddism," our closest word:** There is also the word "neophobia," but this
      is uncommon and normally used only in technical literature. See, for example,
      Patricia Pliner and Karen Hobden. "Development of a Scale to Measure the Trait
      of Food Neophobia in Humans." *Appetite* 19, no. 2 (October 1992):105–20.

87   **Luddism was, in the words of Thomas Pynchon:** Pynchon, 1984.

87   **Children's is one of America's highest-ranked hospitals:** As of 2012, *U.S. News
      & World Report* has ranked Children's near or at the top of its honor roll for more
      than twenty years. Comarow, 2012.

88   **"A man does not attain the status of Galileo":** From "Velikovsky in Collision,"
      in Gould, 1977.

90   **William Syrotuck analyzed 229 cases:** Syrotuck and Syrotuck, 2000. Cited in
      Gonzales, 2004.

CHAPTER 4: HOW WE SEE

92   **"It contains numerous bacteria":** Warren, 2005.

92   ***Every* patient with a duodenal ulcer:** A "duodenal ulcer" is sometimes known as
      a "peptic ulcer." The "acidic passage" is "the duodenum."

93   **It was eventually given the name *Helicobacter pylori*:** *H. pylori* was known as
      Campylobacter pylori, also "pyloric campylobacter," for some years—*H. pylori* is its
      final and current name.

93   **the *Lancet,* one of the world's highest-impact medical journals:** In the 2011
      Journal Citation Report: Science Edition (Thompson Reuters, 2012), the *Lancet*'s
      impact factor was ranked second among general medical journals, at 38.278, after
      the *New England Journal of Medicine,* at 53.298. From Wikipedia's entry on the
      Lancet, at http://bit.ly/lancetwiki.

93   **"appeared to be a new species":** Marshall and Warren, 1984.

93 **Ian Munro was no ordinary journal editor:** Freeman, 1997.

93 **even adding a note saying:** Munro, 1984. Quoted in Van Der Weyden, 2005.

93 **We now know that there are hundreds of species of bacteria:** See, for example, Sheh, 2013.

94 **"As my knowledge of medicine and then pathology increased":** Warren, 2005.

94 **"I preferred to believe my eyes":** Marshall, 2002. Cited in Pincock, 2005.

94 **a group of American scientists:** Ramsey et al., 1979. Six scientists, variously from the University of Texas, Harvard Medical School, and Stanford University, authored the paper.

95 **they were led by a decorated professor of medicine:** John S. Fordtran, who is the last-named author on the paper. Biographical details at Boland, 2012.

95 *H. pylori* **was clearly visible:** From Munro, 1985: "That outbreak was in a series of volunteers taking part in a study involving multiple gastric intubations and the cause was then assumed to have been viral. However, biopsy specimens have now been examined retrospectively and pyloric campylobacters have been found."

95 **"Failing to discover** *H. pylori* **was my biggest mistake":** W. I. Peterson, in a GastroHep.com profile: "What is the biggest mistake that you have made? Failing to discover *H. pylori* in 1976." Available at http://bit.ly/walterpeterson.

95 **In 1967, Susumo Ito, a professor at Harvard Medical School:** Ito, 1967. Cited in Marshall, 2005.

95 **In 1940, Harvard researcher Stone Freedberg:** Freedberg and Barron, 1940. Cited in Marshall, 2005: "The new spiral organism was not just a strange infection occurring in Western Australia, but was the same as the 'spirochaete' which had been described in the literature several times in the previous 100 years. . . . In 1940, Stone Freedberg from Harvard Medical School had seen spirochaetes in 40% of patients undergoing stomach resection for ulcers or cancer. About 10 years later, the leading US gastroenterologist, Eddie Palmer at Walter Reid [*sic*] Hospital, had performed blind suction biopsies on more than 1000 patients but had been unable to find the bacteria. His report concluded that bacteria did not exist except as post mortem contaminants." See also Altman, 2005.

95 *H. pylori* **has now been found in medical literature:** See Kidd and Modlin, 1998; Unge, 2002; and Marshall, 2002.

95 **"inattentional blindness":** Mack and Rock, 2000.

95 **"Something that we can't see, or don't see":** Adams, 2008. The quotation consists of two separate elements edited and combined: "An S.E.P.," he said, "is something that we can't see, or don't see, or our brain doesn't let us see, because we think that it's somebody else's problem. That's what S.E.P. means. Somebody Else's Problem. The brain just edits it out; it's like a blind spot. If you look at it directly you won't see it unless you know precisely what it is. Your only hope is to catch it by surprise out of the corner of your eye," and, later, "The Somebody Else's Problem field is much simpler and more effective, and what is more can be run for over a hundred years on a single flashlight battery. This is because it relies on people's natural predisposition not to see anything they don't want to, weren't expecting or can't explain."

96 **The path from eye to mind is long:** Description of visual loop based on Seger, 2008.

96 **This is why it is a bad idea:** The literature on this point is unequivocal: see, for

example: Harbluk et al., 2002; Strayer et al., 2003; Rakauskas et al., 2004; Strayer and Drews, 2004; Strayer et al., 2006; Strayer and Drews, 2007; and Young et al., 2007.

97 **In one study, researchers put a clown on a unicycle:** Hyman et al., 2010.

97 **Harvard researchers Trafton Drew and Jeremy Wolfe:** Drew et al., 2013, "The Invisible Gorilla Strikes Again."

97 **In 2004, a forty-three-year-old woman:** Lum et al., 2005. The incident took place at Strong Memorial Hospital in Rochester, New York. Cited in Drew et al., 2013.

98 **When Robin Warren accepted his Nobel Prize:** Warren, 2005, citing Doyle, 2011, from "The Boscombe Valley Mystery," first published in 1891.

98 **They can diagnose a disease after looking at a chest X-ray:** Drew et al., 2013.

99 **Adriaan de Groot, a chess master and psychologist:** De Groot, 1978. Cited in Weisberg, 1986.

103 **In 1960, twelve elderly Japanese Americans:** Biographical details about Shun-ryu Suzuki are from Chadwick, 2000.

103 **these men and women were imprisoned:** Americans of Japanese descent living in San Francisco were interred at Tanforan Racetrack, now a shopping mall at 1150 El Camino Real, San Bruno, California, where they were housed in stables and barracks before being moved to other camps farther inland. University of Southern California, 1942; *San Francisco Chronicle,* 1942.

103 **They were Zen Buddhists and congregants of Sokoji:** Chadwick, 2000: "The name he gave the abandoned synagogue had a simple meaning: *Soko* stood for San Francisco and the *ji* meant temple." The original temple was at 1881 Bush Street, four miles southeast of Fort Point and the southern end of the Golden Gate Bridge. The building was originally the Ohabai Shalome synagogue of the Jewish Congregation Ohabai Shalome; it was sold to Japanese American Teruro Kasuga in 1934 after the congregation experienced misfortunes, including a loss of membership due to religious reforms and the murder of its rabbi during what may have been a homosexual encounter. Kasuga turned it into *Sokoji,* also known as the "Soto Zen Center." The congregation moved to larger facilities on Page Street between 1969 and 1972, partly as a result of the increased interest in Zen Buddhism that Shunryu Suzuki had helped create. The building's history is beautifully described in Kenning, 2010.

103 **As the sun rose:** Suzuki arrived on May 23, 1959. Sunrise that day was at around 5:55 a.m. (see http://bit.ly/sfsunrise). Japan Air Lines flight 706 arrived at 6:30 a.m. (http://bit.ly/jaltime). The plane was a DC-6B, with silver-and-white livery, as shown at http://bit.ly/jaldc6; also http://bit.ly/jaldc6b. The "Pacific Courier" designation is from http://bit.ly/jaltime. Suzuki's clothing is described in Chadwick, 2000: "He was wearing his priest's traveling robes with a *rakusu* hanging around his neck, *zori,* and white *tabi* socks."

103 **"I sit at 5:45 in the morning":** Chadwick, 2000.

103 **People in India and East Asia:** From Wikipedia, http://bit.ly/easia: "The UN subregion of Eastern Asia and other common definitions of East Asia contain the entirety of the People's Republic of China, Japan, North Korea, South Korea, Mongolia and Taiwan." According to Everly and Lating, 2002, meditation has been practiced since 1500 B.C.E. Writer Alan W. Watts helped introduce medita-

tion to the United States in the 1959 as the presenter of KQED San Francisco's public television series *Eastern Wisdom and Modern Life*. His episode on meditation, "The Silent Mind," is at http://bit.ly/wattsmind.

103 **Suzuki made his students sit on the floor:** Chadwick, 2000. Picture at http://bit.ly/shunryu.

103 **If he suspected they were sleeping:** The name of the stick is typically transliterated as *keisaku,* but it is called *kyōsaku* in the Soto school, of which Suzuki was a member. Picture at http://bit.ly/kyosaku.

104 **His was American Buddhism's first voice:** Suzuki, 1970. From Fields, 1992: "It was, in fact, an American Buddhist voice, unlike any heard before, and yet utterly familiar. When Suzuki Roshi spoke, it was as if American Buddhists could hear themselves perhaps for the first time." Fields is cited in the 2011 edition of Suzuki, *Zen Mind, Beginner's Mind.*

104 **Nyogen Senzaki, one of the first Zen monks in America:** Senzaki, 1919.

104 **David Foster Wallace made the same point:** Wallace, 2009.

105 **Kuhn was recovering from a great disappointment:** Biographical details about Thomas Kuhn are from Nickles, 2002.

106 **This change in Kuhn's path:** Kuhn, 1977: "One memorable (and very hot) summer day those perplexities suddenly vanished." Nickles, 2002, citing Caneva, 2000, quotes Kuhn describing the event as taking place during an "afternoon," while attending a ceremony at the University of Padua, Italy, in 1992. Weinberg, 1998, also describes talking with Kuhn at this event about his understanding of Aristotle.

106 **The conventional view was that the book:** See, for example, Heidegger, 1956: "Aristotelian 'physics' . . . determines the warp and woof of the whole of Western thinking, even at that place where it, as modern thinking, appears to think at odds with ancient thinking. But opposition is invariably comprised of a decisive, and often even perilous, dependence. Without Aristotle's Physics there would have been no Galileo." Cited in the Wikipedia entry on Aristotle's *Physics,* at http://bit.ly/aristotlephysics.

106 **"Everything that is in locomotion":** Edited from Aristotle, 2012. The complete quotation is: "Everything that is in locomotion is moved either by itself or by something else. In the case of things that are moved by themselves it is evident that the moved and the movement are together: for they contain within themselves their first movement, so that there is nothing in between. The motion of things that are moved by something else must proceed in one of four ways for there are four kinds of locomotion caused by something other than that which is in motion, viz.: pulling, pushing, carrying, and twirling. All forms of locomotion are reducible to these."

106 **Science is not a continuum, he concluded:** Another example, discussed at length by Kuhn, 1962: in 1667, German Johann Joachim Becher published a book called *Physical Education,* in which he first described his theory of how and why things burned. Becher identified a new element called *"terra pinguis,"* which was a part of anything that burned. Burning released *terra pinguis* into the air until the air was so full of *terra pinguis* that it could take no more, at which point the burning stopped. Things that did not burn contained no *terra pinguis.* In the eighteenth century, Georg Ernst Stahl changed the name of *terra pinguis* to "phlogiston," and

the theory dominated physics for almost a hundred years. Phlogiston, or *terra pinguis,* has no modern equivalent—according to current science, it does not exist.

107 **Despite its obscure topic, Kuhn's book:** Garfield, 1987: "The 10 most-cited books, in descending order, are Thomas S. Kuhn's *Structure of Scientific Revolutions* . . ." In May 2014, Google Scholar listed more than seventy thousand citations for the book (http://bit.ly/kuhncitations). In 2012, on the fiftieth anniversary of its release, the University of Chicago Press said, "We had no idea that we had a book on our hands that would sell over 1.4 million copies." Press release at http://bit.ly/1pt4million.

107 **"the most influential work of philosophy":** Gleick, 1996. The book started a debate in philosophy that continues today. Critics have accused Kuhn of using "paradigm" to mean many different things (see, for example Masterman, 1970, in Lakatos et al., 1970; Eckberg and Hill, 1979; Fuller, 2001), but they all add up to one thing: a paradigm is a way of seeing the world. The word "paradigm" also became so well known that it appeared in several cartoons in the *New Yorker,* including one in which a doctor tells a patient, "I'm afraid you've had a paradigm shift" (J. C. Duffy, December 17, 2001, at http://bit.ly/paradigmcartoon1) and one in which one unlucky-looking man says to another, "Good news—I hear the paradigm is shifting" (Charles Barsotti, January 19, 2009, at http://bit.ly/paradigmcartoon2).

107 **"During revolutions scientists see new and different things":** Kuhn, 1962. The sudden appearance of *H. pylori* is not a new phenomenon. One example from Kuhn: in 1690, Britain's astronomer royal John Flamsteed saw a star and called it "34 Tauri." In 1781, William Herschel looked at it through a telescope but saw a comet, not a star. He pointed it out to Nevil Maskelyne, who saw a comet that might be a planet. German Johann Elert Bode saw a planet, too, and this soon led to a consensus: the object was a planet and was eventually called Uranus. Once one new planet had been discovered, the paradigm changed: finding new planets seemed possible. Astronomers, using the same instruments as before to look at the same sky, suddenly found twenty more minor planets and asteroids, including Neptune, which, like Uranus, had looked like a star since the seventeenth century. Something similar happened when Copernicus said the earth revolved around the sun: the previously unchangeable sky suddenly filled with comets that had been made visible not by new instruments but by a new paradigm. Meanwhile, Chinese astronomers, who had never believed that the sky was unchangeable, had been seeing comets for centuries.

108 **Neil deGrasse Tyson, speaking at the Salk Institute:** Tyson, 2006. Video at http://bit.ly/NdGTSalk. Quotation edited from the transcript at http://bit.ly/NdGTsenses. The complete quotation as transcribed is: "And we so much praise about the human eye, but anyone who has seen the full breadth of the electromagnetic spectrum will recognize how blind we are, okay, and part of that blindness means we can't see, we can't detect, magnetic fields, ionizing radiation, radon. We are like sitting ducks for ionizing radiation. We have to eat constantly, because we're warm blooded. Crocodile eat a chicken a month, it's fine. Okay, so we are always looking for food. These gases at the bottom [referring to a slide, with the words CO (carbon monoxide), CH_4 (methane), CO_2 (carbon dioxide)]: you can't smell them, taste them you breath [*sic*] them in you're dead, okay."

109 **"After work you have to get in your car"**: Heavily edited for length from Wallace, 2009.

110 **the original Chinese idea of yin-yang**: In simplified Chinese: 阴阳; traditional Chinese: 陰陽. The characters mean "sunny-side, shady-side." There is no "and."

111 **"an investigation into the condition"**: Lowell's comments are edited from a quotation in Sheehan, 1996, which cites Strauss, 1994. The original quotation from Sheehan is: "What Percival Lowell hoped to accomplish through this 'speculative, highly sensational and idiosyncratic project' is well documented in an address he gave to the Boston Scientific Society on May 22, 1894, which was printed in the *Boston Commonwealth*. His main object, he stated, was to study the solar system: 'This may be put popularly as an investigation into the condition of life on other worlds, including last but not least their habitability by beings like [or] unlike man. This is not the chimerical search some may suppose. On the contrary, there is strong reason to believe that we are on the eve of pretty definite discovery in the matter.' To Lowell, the implications of the lines that Italian astronomer Giovanni Schiaparelli figuratively called *canali* were self-evident: 'Speculation has been singularly fruitful as to what these markings on our next to nearest neighbor in space may mean. Each astronomer holds a different pet theory on the subject, and pooh-poohs those of all the others. Nevertheless, the most self-evident explanation from the markings themselves is probably the true one; namely, that in them we are looking upon the result of the work of some sort of intelligent beings. . . . The amazing blue network on Mars hints that one planet besides our own is actually inhabited now.'" Sheehan's work is available from the University of Arizona at http://bit.ly/sheehanmars.

111 **Lowell inspired a century of science fiction**: The word "Martian" predated Lowell—it first appeared in 1883, in a story almost certainly inspired by Schiaparelli (Lach-Szyrma, 1883) but did not become famous until 1898, *after* Lowell's announcements, when H. G. Wells published *The War of the Worlds*. Burroughs's *Under the Moons of Mars* was a series of short stories first published in 1912, as a series under the pen name "Norman Bean," and renamed *A Princess of Mars* when released in book form (Burroughs, 1917). The complete quotation is: "The shores of the ancient seas were dotted with just such cities, and lesser ones, in diminishing numbers, were to be found converging toward the center of the oceans, as the people had found it necessary to follow the receding waters until necessity had forced upon them their ultimate salvation, the so-called Martian canals."

111 **One of Lowell's opponents was Alfred Wallace**: Wallace had already concluded that "the Earth is the only habitable planet in the solar system" when Lowell started publishing (Wallace, 1904).

111 **"The totally inadequate water-supply"**: Edited from Wallace, 1907.

112 **The argument was resolved in Wallace's favor**: Momsen, 1996. The complete quotation is: "And then the real wonder came—picture after picture showing that the surface was dotted with craters! It appeared uncannily like that of our own Moon, deeply cratered, and unchanged over time. No water, no canals, no life." Momsen was described as "the imaging engineer for JPL's [Jet Propulsion Laboratory's] Mariner series of missions" by John B. Dobbins on December 12, 2005, in a message to the NASA Spaceflight Forum at http://bit.ly/nasaforum.

112 **His maps of Martian canals are mirror images:** Sheehan and Dobbins, 2003. Lowell describes the "Tores" he saw on Saturn in Lowell, 1907.

113 **"Perhaps the most harmful imperfection of the eye":** See Sheehan and Dobbins, 2003; also Douglass, 1907.

113 **"I am not sure of the significance":** Warren, 2005.

113–114 **Ketamine, phencyclidine, and methamphetamine:** Burton, 2009. Phencyclidine is also known as PCP, or angel dust. Methamphetamine is also known as "meth"; the derivatives MDMA, or ecstasy, and methamphetamine hydrochloride salt, or "crystal meth," can also create feelings of certainty. For more on the effects of entorhinal cortex stimulation, see Bartolomei, 2004.

114 **cognitive psychologists Ulric Neisser and Nicole Harsch:** Neisser and Harsch, 1992. Cited in Burton, 2009.

114 **Thirty-three were sure they had never been asked:** This was actually thirty-three out of forty-four. The study had three parts. In part one, 106 students completed a questionnaire the day after the *Challenger* explosion. In part two, administered two and a half years later, forty-four of those students agreed to complete a follow-up questionnaire. In part three, forty of those students participated in an interview where the two questionnaires were compared. Part three, the interview, took place six months after part two, the second questionnaire, had been completed. Four students dropped out between the second and third parts of the test, which is why the base size in the test is forty.

114 **This unshakable certainty was first studied in 1954:** Festinger et al., 1956, in which Martin is given the pseudonym "Mrs. Marian Keech" to protect her identity.

115 **"The group began reexamining the original message":** Edited from Festinger et al., 1956. Complete quotation: "At any rate, in the next hour and a half, the group began to come to grips with the fact that no caller had arrived at midnight to take them to the saucer. The problem from here on was to reassure themselves and to find an adequate, satisfying way to reconcile the disconfirmation with their beliefs. They began by re-examining the original message which had stated that at midnight the group would be put into parked cars and taken to the saucer. In response to some of the observers' prodding about that message during the coffee break, the Creator stated that anyone who wished might look up that message. It had been buried away among many others in a large envelope and none of the believers seemed inclined to look for it, but one of the observers volunteered. He found it and read it aloud to the group. The first attempt at reinterpretation came quickly. Daisy Armstrong pointed out that the message must, of course, be symbolic, because it said we were to be put into parked cars; but parked cars do not move and hence could not take the group anywhere. The Creator then announced that the message was indeed symbolic, but the 'parked cars' referred to their own physical bodies, which had obviously been there at midnight. The 'porch' (flying saucer), He went on, symbolized in this message the inner strength, the inner knowing, and inner light which each member of the group had. So eager was the group for an explanation of any kind that many actually began to accept this one."

115 **"From the mouth of death have ye been delivered":** Edited from Festinger et al., 1956. Complete quotation: "And mighty is the word of God—and by his word have ye been saved—for from the mouth of death have ye been delivered

and at no time has there been such a force loosed upon the Earth. Not since the beginning of time upon this Earth has there been such a force of Good and light as now floods this room and that which has been loosed within this room now floods the entire Earth."

116 **Leon Festinger, named this gap:** The term used throughout *When Prophecy Fails* is "dissonance." Later, in Festinger, 1957, the term became "cognitive dissonance."

115 **In one experiment, he gave volunteers:** Festinger, 1962.

116 **"When dissonance is present, in addition to trying to reduce it":** Festinger, 1957.

116 **Dorothy Martin had a long career:** After the events described in the book, Martin moved to the Yucatán Peninsula in Mexico, was involved with "the Brotherhood of the Seven Rays," a group that included another purported "UFO contactee," George Hunt Williamson, and at some point became known as "Sister Thedra." According to another spiritualist, "Dr. Robert Ghost Wolf," while she was in Mexico, Martin "had an experience which changed her in an instant when as it is told by her that [*sic*] Jesus Christ physically appeared to her and spontaneously cured her of cancer. He introduced himself to her by his true, [*sic*] name, 'Sananda Kumara,' thereby revealing his affiliation with the Venusian founders of the Great Solar Brotherhoods. By his command that [*sic*] Sister Thedra went to Peru. Sister Thedra eventually left Peru upon felling [*sic*] her experience there was complete. She then traveled to Mt. Shasta in California and founded the Association of Sananda and Sanat Kumara." Dorothy Martin died in May 1992. She did her last "automatic writing" on May 3, 1992: "Sori Sori: Mine beloved, I am speaking unto thee for the good of all. It is now come the time that ye come out from the place wherein ye are. Ye shall shout for joy! Let it be, for many shall greet thee with glad shouts! So be it, no more pain . . . Amen . . . Sananda." (Ellipses in original.) After Martin's death, the Association of Sananda and Sanat Kumara changed its address to a location next to a pizza restaurant called "Apizza Heaven" in Sedona, Arizona. See http://bit.ly/thedra and http://bit.ly/sananda. Martin's story is also mentioned (rather inaccurately) in Largo, 2010.

116 **"The psychologists determined that when people":** From "Extraordinary Intelligence," a website created by a woman using the pseudonym "Natalina," sometimes "Natalina EI," who lives in Tulsa, Oklahoma; http://bit.ly/whenfaithistested.

CHAPTER 5: WHERE CREDIT IS DUE

118 **Sleet like crystal tears fell on cobbles:** Biographical details about Rosalind Franklin are from Maddox, 2003, and Glynn, 2012.

118 **Physicist Erwin Schrödinger captured the spirit:** Schrödinger gave a series of lectures at the Dublin Institute for Advanced Studies at Trinity College in 1943 (published as a book in 1944) in which he anticipated the discovery of DNA with the statement "the most essential part of a living cell—the chromosome fibre may suitably be called an aperiodic crystal" (Schrödinger, 1944).

120 **Mendel's work was ignored:** Mendel's work did not become widely known until the start of the twentieth century; Darwin died in 1882. Darwin proposed a "pro-

visional hypothesis," quite different from Mendel's, which he called "pangenesis," in Darwin, 1868. "Chromosome theory" is also known as "Boveri-Sutton chromosome theory," "the chromosome theory of inheritance," and "the Sutton-Boveri theory."

120 **Rosalind Franklin believed life's messengers:** It was not until the 1930s that the acids were first considered as candidate information carriers by the Canadian American scientist Oswald Avery Jr. (Maddox, 2003).

121 **a crystal is any solid with atoms or molecules arranged:** A crystal can also consist of a three-dimensional, repeating arrangement of ions; I excluded that point here for clarity and simplicity.

122 **Franklin published her results at the start:** Franklin published regularly on the tobacco mosaic virus between 1955 and 1958 (see works by Franklin and by Franklin with others in the bibliography, below), and her work culminated in two papers published in 1958: "The Radial Density Distribution in Some Strains of Tobacco Mosaic Virus," coauthored with Kenneth Holmes and published before her death (Holmes and Franklin, 1958), and "The Structure of Viruses as Determined by X-ray Diffraction," which was published posthumously (Franklin et al., 1958).

123 **"Credit does not entirely belong to her":** A letter from Charles Eliot to Marie Meloney, December 18, 1920, part of the Marie Mattingly Meloney papers, 1891–1943, Columbia University Library; http://bit.ly/meloney. Quoted in Ham, 2002.

124 **Curie used the word "me" seven times:** See Curie, 1911. The quotation also appears in Emling, 2013.

124 **In total, only 15 women have won:** "Science" means prizes in "chemistry," "physics," or "physiology or medicine." The 15 women (as of 2014) are Maria Goeppert Mayer (physics, 1963), Marie Curie (physics, 1903, and chemistry, 1911), Ada E. Yonath (chemistry, 2009), Dorothy Hodgkin (chemistry, 1964), Irène Joliot-Curie (chemistry, 1935), Elizabeth H. Blackburn (physiology or medicine, 2009), Carol W. Greider (physiology or medicine, 2009), Françoise Barré-Sinoussi (physiology or medicine, 2008), Linda B. Buck (physiology or medicine, 2004), Christiane Nüsslein-Volhard (physiology or medicine, 1995), Gertrude B. Elion (physiology or medicine, 1998), Rita Levi-Montalcini (physiology or medicine, 1986), Barbara McClintock (physiology or medicine, 1983), Rosalyn Yalow (physiology or medicine, 1977), and Gerty Theresa Cori (physiology or medicine, 1947). See http://bit.ly/womenlaureates.

126 **It protected DNA specimens from humidity:** Pictures of Franklin's camera are at http://bit.ly/dnacamera.

127 **There are many similar stories:** These examples are a selection from Byers and Williams, 2010.

127 **One reason is an imbalance first recorded:** Zuckerman, 1965.

127 **"The world is peculiar in this matter":** Quotations are from Merton, 1968.

127 **Until Zuckerman, most scholars assumed:** See, for example, Pareto et al., 1935, discussed in Zuckerman, 1977.

128 **"For whoever has will be given more":** New International Version. Other translations and commentaries at http://bit.ly/matthew2529.

128 **Zuckerman collaborated with Merton, then married him:** Merton and Zuckerman married in 1993. Merton separated from his first wife, Suzanne Carhart,

in 1968, soon after Zuckerman completed her PhD (Hollander, 2003; Calhoun, 2003; and Wikipedia entry on Robert K. Merton at http://bit.ly/mertonrk).

128 **Patent law is complicated:** See U.S. Patent and Trademark Office web page at http://bit.ly/inventorship.

128 **If the female scientist named the male scientist:** See Radack, 1994, for a discussion of the risks of assigning inventorship to non-inventors.

129 **the average number of people who "contribute":** See discussion of Trajtenberg in chapter 1.

129 **part of the macroenvironment:** Merton used the word "paradigm" twenty-five years before Kuhn, but, Merton says, with a less precise, "more limited," meaning. See video of "Robert K. Merton Interviewed by Albert K. Cohen, May 15, 1997," posted by the American Society of Criminology at http://bit.ly/mertoncohen.

130 **In 1676, Isaac Newton described this problem:** Letter from Isaac Newton to Robert Hooke, dated "Cambridge, February 5, 1675–6," published in Brewster, 1860.

130 **Newton got it from George Herbert:** See Merton, 1993. There is an excellent summary of the life of this quotation, written by Joseph Yoon, formerly of NASA, on Aerospace Web at http://bit.ly/josephyoon (although the date given for Didacus Stella's quotation is incorrect). Bernard of Chartres may have found the idea in the work of Talmudic scholars (it appears in the writings of Talmudist Isaiah di Trani, who lived after Bernard, but Isaiah may have inherited it from other, earlier Talmudists, rather than getting it from Bernard); it could also have been inspired by the ancient Greek myth of Cedalion, who rides on the shoulders of the giant Orion.

131 **a subject of curiosity at least since the winter:** There are other, earlier discussions of snowflakes, including Han Ying (韓嬰, 150 B.C.E.), Albertus Magnus (1250), and Olaus Magnus (1555). I start with Kepler because he was one of the first to try to explain snowflakes by connecting them to crystals—"Let the chemists, then, tell us whether there is any salt in snow, and what kind, and what shape it takes"—and crystals, not snowflakes, are the subject of the discussion.

132 **Geissler's invention was a novelty:** Shepardson, 1908.

132 **"I have seen my death":** Markel, 2012.

133 **Were they particles, like electrons:** This question about X-rays was asked before Einstein proposed wave-particle duality.

133 **In 1915, at the age of twenty-five:** Jenkin, 2008, and Authier, 2013.

134 **One of them was a woman named Polly Porter:** Polly was not her real name. She was christened "Mary Winearls Porter" but had always been called "Polly."

134 **While her brothers studied, Porter wandered the city:** The result was a book, *What Rome Was Built With.* See Porter, 1907.

134 **Henry Miers, Oxford's first professor of mineralogy:** Price, 2012.

135 **"Dear Professor Goldschmidt":** Letter dated January 14, 1914, edited from Arnold, 1993. The quotation as it appears in Arnold is: "Dear Professor Goldschmidt: I have long had the purpose of writing you to interest you in Miss Porter, who is working this year in my laboratory and whom I hope you will welcome in your laboratory next year. Her heart is set upon the study of crystallography and I hope she will remain with you for more than one year. Her income is not sufficient for her to live in Bryn Mawr College without earning money. This

Miss Porter is doing now, but her work takes too much time from her studies and besides she should go to the fountainhead of inspiration. . . . Miss Porter thinks she will, in Germany, be able to live upon her income. Miss Porter's life has been unusual, for her parents (her father is corresponding editor of the London Times) have been almost constant travellers and she has never been to school or college save for a very brief period. There are therefore great gaps in her education, particularly in chemistry and mathematics, but to offset this I believe you will find that she has an unusual aptitude for crystal measurement, etc., and certainly an intense love of your subject. I want to see her have the opportunities which have so long been denied her—Miss Porter is perhaps about 26 years of age, very modest and unselfassertive but with a quiet initiative. I hope you will be interested to have her as a student and I think she will repay all you may do for her. She must eventually be self-supporting and I hope she will be fitted for the position of curator and crystallographer of some mineral collection. Miss Porter is spending this year only with me and if she does come to you, it will be apparent to you, I fear, that she has but made a beginning. I am, however, both ambitious for her and with faith in her ultimate success. . . ."

135 **She stayed at Oxford, conducting research:** Haines, 2001.

136 **Bragg's topic in 1923:** The title of Bragg's lectures was "Concerning the Nature of Things." Bragg, 1925.

136 **Dorothy Hodgkin:** Biographical details about Dorothy Hodgkin are from Ferry, 2000.

137 **That same year, Japanese physicist Ukichiro Nakaya:** Nakaya, 1954. Summarized in nontechnical terms in Libbrecht, 2001.

137 **They form around another particle:** See Lee, 1995, for more. Christner et al., 2008, "examined IN [ice nucleators—particles that act as a nucleus for ice crystals that form in the atmosphere] in snowfall from mid- and high-latitude locations and found that the most active were biological in origin. Of the IN larger than 0.2 micrometer that were active at temperatures warmer than -7°C, 69 to 100% were biological, and a substantial fraction were bacteria."

137 **Nucleobases, essential components of DNA:** Callahan et al., 2011.

137 **glycolaldehyde, a sugarlike molecule:** Jørgensen et al., 2012.

138 **Franklin likely inherited:** Gabai-Kapara, 2014, suggests that only 2 percent of Ashkenazi Jews carry a BRCA mutation, split evenly between the BRCA1 mutation and the BRCA2 mutation. (Only about three in ten thousand Ashkenazim have mutations in both their BRCA1 *and* BRCA2 genes.) Not all women with BRCA mutations develop ovarian cancer, and not all ovarian cancers among Ashkenazi Jewish women are caused by BRCA mutations: only 40 percent of Ashkenazi Jewish women who develop ovarian cancer have BRCA2 mutations. It is Franklin's death from ovarian cancer at such a young age, *combined with* her Ashkenazi Jewish descent, that indicates she was likely to have been a carrier of a mutated BRCA gene.

138 **The BRCA2 mutation makes:** Antinou, 2003. While 1.4 percent of all women develop ovarian cancer, 39 percent of women with a BRCA1 mutation and 11 to 17 percent of women with a BRCA2 mutation develop ovarian cancer. BRCA mutations also increase breast cancer risk: while 12 percent of all women develop breast cancer, 55 to 65 percent of women with a BRCA1 mutation and 45 percent

of women with a BRCA2 mutation develop breast cancer. See the National
Cancer Institute at http://bit.ly/ncibrca for more information about the impact
of BRCA mutations on both diseases.

138 **all literal cousins of Rosalind Franklin:** According to genetic analysis by Carmi,
2014, all Ashkenazi Jews are descended from a population of about 350 people
who lived seven hundred years ago, around 1300 CE. If we assume a generation is,
on average, twenty-five years, and the founding people were interrelated, this suggests that all living Ashkenazim are about thirtieth cousins or closer.

CHAPTER 6: CHAINS OF CONSEQUENCE

140 **William Cartwright's dog started barking:** Details of the attack on Cartwright's mill taken from the "Luddite Bicentenary" website at: http://bit.ly
/rawfolds.

141 **The new and improved Enoch sledgehammers:** Details about "the Great
Enoch" are available on the *Radical History Network* blog at http://bit.ly
/greatenoch.

142 **"Governments must have arisen":** Paine, 1791.

146 **He begins with Frenchman Philippe Lebon:** Lebon's patent is dated 1801, but
Ehrenburg describes him developing the engine in 1798 (Ehrenburg, 1929).

147 **"It really boils down to this":** Dr. King first delivered this sermon at Ebenezer
Baptist Church, where he served as co-pastor. On Christmas Eve 1967, the Cana-
dian Broadcasting Corporation aired the sermon as part of the seventh annual
Massey Lectures. Available at http://bit.ly/drkingsermon.

149 **"We do not consider modern inventions to be evil":** This quotation, and other
details about the Amish, are from Kraybill et al., 2013.

150 **"Not everything that could be fixed should be fixed":** Morozov, 2013.

156 **When all the processes in Coca-Cola's tool chain:** Analysis based on Ercin et
al., 2011.

158 *Come on, my love:* This is the English translation of a traditional Scottish walking, or "waulking," song "Coisich, A Ruin" ("Come On, My Love"), probably
from around the fourteenth century. There is a beautiful recording by Catriona
MacDonald at http://bit.ly/coisich. Craig Coburn summarizes the tradition of the
Scottish walking song at http://bit.ly/craigcoburn. Fulling in England is discussed
in Pelham, 1944; Lennard, 1951; Munro, 1999; and Lucas, 2006.

160 **jobs that, less than a century later:** Dating this to Towne, 1886.

161 **Between 1840 and 1895, school attendance:** Cipolla, 1969.

161 **In 1990, America had 30 million:** Statistics from Snyder, 1993, summarized at
http://bit.ly/snydersummary; full version at http://bit.ly/snyderthomas.

161 **The number of Americans earning college degrees:** Analysis based on
demographic data from InfoPlease, "Population Distribution by Age, Race, and
Nativity, 1860–2010" (http://bit.ly/uspopulation); U.S. Census at http://bit.ly
/educationfacts; Snyder, 1993 (http://bit.ly/snyderthomas); and Joseph Kish's table
"U.S. Population 1776 to Present" (http://bit.ly/kishjoseph).

163 **In March 2002, Woody Allen did something:** Biographical details about Woody Allen from Wikipedia entry at http://bit.ly/allenwoody. In 2002 he had won three Academy Awards—two for *Annie Hall* (Best Original Screenplay and Best Director, 1978) and one for *Hannah and Her Sisters* (Best Original Screenplay, 1987). He had also been nominated for seventeen other awards: *Annie Hall* (Best Actor in a Leading Role, 1978), *Interiors* (Best Original Screenplay and Best Director, 1979), *Manhattan* (Best Original Screenplay, 1980), *Broadway Danny Rose* (Best Original Screenplay and Best Director, 1985), *The Purple Rose of Cairo* (Best Original Screenplay, 1986), *Hannah and Her Sisters* (Best Director, 1987), *Radio Days* (Best Original Screenplay, 1988), *Crimes and Misdemeanors* (Best Original Screenplay and Best Director, 1989), *Alice* (Best Original Screenplay, 1990), *Husbands and Wives* (Best Original Screenplay, 1993), *Bullets Over Broadway* (Best Original Screenplay and Best Director, 1994), *Mighty Aphrodite* (Best Original Screenplay, 1996), and *Deconstructing Harry* (Best Original Screenplay, 1998). As of 2014, since appearing at the 2002 Academy Awards ceremony, he has won a fourth award for *Midnight in Paris* (Best Original Screenplay, 2011) and received three other nominations: *Match Point* (Best Original Screenplay, 2006), *Midnight in Paris* (Best Director, 2011), and *Blue Jasmine* (Best Original Screenplay, 2014). A complete list of Allen's awards is available at the Internet Movie Database, http://bit.ly /allenawards. The speech in which he said, "For New York City, I'll do anything" can be seen on YouTube at http://bit.ly/allenspeech.

163 **He gives several tongue-in-cheek excuses:** From Block and Cornish, 2012: "Audie Cornish, Host: Woody Allen is a favorite to take home at least one Oscar for Best Original Screenplay, but don't expect the camera to cut to him when the nominees' names are announced. Melissa Block, Host: With one exception, Woody Allen has never attended the Academy Awards. In spite of his previous 21 nominations and three wins, he declines the invites. He's known for it, so notoriously so that urban myths are told as to why. Cornish: No, it's not because of a standing gig playing the clarinet at a New York pub. We were assured of that by Eric Lax, who wrote 'Conversations with Woody Allen.' Eric Lax: It was a polite excuse. I think that, if he has a gig that night, he can say, well, I had a gig that night. I needed to be there. You know, that goes all the way back to 'Annie Hall.'" Allen, quoted in Hornaday, 2012: "They always have it on Sunday night. And it's always—you can look this up—it's always opposite a good basketball game. And I'm a big basketball fan. So it's a great pleasure for me to come home and get into bed and watch a basketball game. And that's exactly where I was, watching the game."

164 **"The whole concept of awards is silly":** From Lax, 2000: "There are two things that bother me about [the Academy Awards]," he said in 1974 after Vincent Canby had written a piece wondering why *Sleeper* had received no nominations. "They're political and bought and negotiated for—although many worthy people have deservedly won—and the whole concept of awards is silly. I cannot abide by the judgment of other people, because if you accept it when they say you deserve an award, then you have to accept it when they say you don't."

164 **"I think what you get in awards is favoritism":** From Weide, 2011. Video clip on YouTube at http://bit.ly/whatyougetinawards.

164　**Psychologist R. A. Ochse lists eight motivations:** Ochse, 1990.

164　**"I want to feel my work good and well taken":** Plath, 1982, as quoted in Amabile, 1996.

165　**Amabile asked ninety-five people:** Amabile, 1996.

166　**Olympia SM2 portable typewriter:** Australian blogger Teeritz gives a detailed description of the SM2, with photographs, at http://bit.ly/olympiasm2.

166　**"It still works like a tank":** Woody Allen quotations throughout from Lax, 2000, and Weide, 2011; descriptions (e.g., type of typewriter) based on Weide, 2011.

167　**Poet John Berryman congratulated him:** Simpson, 1982. Cited in Amabile 1983.

167　**"When I began to think of what":** Edited from Eliot, 1948. Full text at http://bit.ly/eliotbanquet.

168　**an address to the Nordic Assembly of Naturalists:** Einstein, 1923.

168　**The days are cold and dry:** Sausalito weather in February 1976 from *Old Farmer's Almanac* at http://bit.ly/pointbonita.

168　**A strange redwood hut:** Record Plant Studios, 2200 Bridgeway, Sausalito, CA 94965. Photographs of the entrance, with carved animals, at http://bit.ly /recordplant.

169　**Christine McVie calls it a "a cocktail party":** Crowe, 1977. Complete quotation: "'Trauma,' Christine groans. 'Trau-*ma*. The sessions were like a cocktail party every night—people everywhere.'"

169　**Warner Bros. compares it to the rocket:** *Tusk* has a mixed reputation now. Some critics, and some members of Fleetwood Mac, regard it as the band's best work.

170　***Don't Stand Me Down* confused reviewers:** As with *Tusk,* some now consider *Don't Stand Me Down* to be a misunderstood masterpiece. See, for example, comments on the *Guardian* website at http://bit.ly/dontstand, such as, "*Don't Stand Me Down* is the statement of a maverick genius that went over the heads of all but the connoisseurs."

170　**Dexys Midnight Runners would not record:** Details about Dexys Midnight Runners and *Don't Stand Me Down* at Wikipedia: http://bit.ly/dexyswiki and http://bit.ly/dontstandwiki. General discussion of "second album syndrome" in Seale, 2012.

170　**"This is my story":** Dostoyevsky, 1923; partly quoted in Amabile 1983, citing Allen, 1948.

171　**After getting a doctorate in psychology:** Biographical details about Harry Harlow are from Sidowski and Lindsley, 1988, and the Wikipedia entry for Harry Harlow at http://bit.ly/harlowharry.

171　**Harlow left puzzles consisting of a hinge:** See Harlow, 1950.

172　**"tended to disrupt, not facilitate":** Harlow et al., 1950.

172　**They consistently rated the commissioned art:** Amabile, Phillips, and Collins, 1994, cited in Amabile, "Creativity in Context," 1996.

172　**Princeton's Sam Glucksberg investigated the question:** Glucksberg, 1962, cited in Amabile, 1983.

172　**Follow-up experiments by Glucksberg:** For example, McGraw and McCullers, 1979.

172　**There are more than a hundred studies:** See for example, reviews by Cameron and Pierce, 1994; Eisenberger and Cameron, 1996; and Eisenberger et al., 1999.

173 **Rewards are only a problem:** McGraw and McCullers, 1979, cited in Amabile 1983.

173 **Amabile explored and extended this finding:** Amabile,Hennessey, and Gross-man (1986), cited in Amabile, "Creativity in Context," 1996.

174 **People in America's Deep South:** Among many excellent books about Robert Johnson are Wardlow, 1998; Pearson and McCulloch, 2003; and Wald, 2004.

177 **One has even attributed it to "cramping":** Flaherty, 2005.

177 **He wrote a play called *Writer's Block*:** *Writer's Block* is two one-act plays. The description given in many playbills is: "In *Riverside Drive,* a paranoid schizo-phrenic former-screenwriter stalks a newly successful but insecure screenwriter, believing he has stealing not only his ideas but his life. *Old Saybrooke,* a combi-nation of old-fashioned sex farce and an interesting look at a writer's process, involves a group of married couples who have cause to ponder the challenges of commitment." See, for example, Theatre in LA at http://bit.ly/theatreinla and Goldstar at http://bit.ly/goldstarhollywood.

177 **"For the first time in my life":** Transcript of *Deconstructing Harry* corrected from Drew's Script-O-Rama at http://bit.ly/harryblock.

177 **Allen took the role of Harry:** Details about Allen's process, and Allen quota-tions, from Lax, 2000.

179 **"You have to dip your pen in blood":** Allen may be thinking of the following comment, reported by pianist Alexander Goldenveizer in a memoir translated by S. S. KotelDiansky and Virginia Woolf as *Talks with Tolstoi,* published by the Hogarth Press in 1923: "One ought only to write when one leaves a piece of one's flesh in the ink-pot each time one dips one's pen." This extract from the memoir is also referenced in Walter Allen's *Writers on Writing,* 1948.

181 *Popular Science* **described them as "savages":** Barrows, 1910.

181 **The last anthropologist to live:** *Boston Evening Transcript,* 1909.

181 **Rosaldo captured the Ilongots' insights in a book:** Rosaldo, 1980.

182 **"The force of any passion or emotion":** Spinoza, 1677.

182 **"We can't be misled by passions":** Descartes, 1649.

182 **Daquan Lawrence celebrated his sixteenth birthday:** Daquan Lawrence's story and lyrics are from Hansen, 2012.

184 **the Irene Taylor Trust claimed:** The claims, which have been repeated in sev-eral of the Trust's publications as well as by other sources, refer to a production of Shakespeare's *Julius Caesar* at Bullingdon Prison in Oxfordshire, England, in May 1999. According to the Trust's original evaluation report, "94% of partici-pants did not offend during the time that they were involved in the Julius Caesar Project" and "There was a 58% decrease in the offence rates of participants in the six months following the project, compared to the offence rates in the six-month period before the project began." The full report, which is called "Julius Caesar—H.M.P Bullingdon," and is undated and attributed only to the "Irene Taylor Trust," can be downloaded from http://bit.ly/taylortrust.

185 **In the 1950s, George "Shotgun" Shuba:** George Shuba story from Kahn, 1972, cited in Glasser, 1977, which mistakenly calls Shuba "Schuba."

185 **what psychologist William Glasser later called:** Glasser, 1976.

186 **"I'll start with scraps and things":** Allen in Weide, 2011.

186 **"To begin, to begin":** From the movie *Adaptation* (2002), directed by Spike Jonze. These lines are written by Charlie Kaufman and said by the character

"Charlie Kaufman," a screen writer struggling with a script, played by Nicholas Cage.

187 **"Work brings inspiration":** Notes about Stravinsky, including this quotation, from Gardner, 2011.

188 **Science describes the destruction unequivocally:** See, for example, Bailey, 2006, which details experimental results and also includes a good literature review.

191 **Woody Allen has pondered that:** Lax, 2010. The complete quotation from Allen is:

"Why not opt for a sensual life instead of a life of grueling work? When you're at heaven's gate, the guy who has spent all his time chasing and catching women and has a sybaritic life gets in, and you get in, too. The only reason I can think of not to is, it's another form of denial of death. You delude yourself that there's a reason to lead a meaningful life, a productive life of work and struggle and perfection of one's profession or art. But the truth is, you could be spending that time indulging yourself—assuming you can afford it—because you both wind up in the same place.

"If I don't like something, it doesn't matter how many awards it's won. It's important to keep your own criteria and not defer to the trends of the marketplace.

"I hope that somewhere along the line it will be perceived that I'm not really a personal malcontent, or that my ambition or my pretensions—which I freely admit to—are not to gain power. I only want to make something that will entertain people, and I'm stretching myself to do it."

CHAPTER 8: CREATING ORGANIZATIONS

193 **In January 1944, Milo Burcham strolled:** Descriptions of the Skunk Works drawn mainly from Johnson, 1990, and Rich, 1994.

193 *Lulu Belle*'s **official name:** To be precise: Lockheed's prototypes, or "experimental" aircraft, had the prefix "X" in their names, so *Lulu Belle*'s full official name was the "XP-80." The P-80 was the name of subsequent production aircraft based on her design.

198 **"When you're dealing with a creative process":** Frank Filipetti quotation from Massey, 2000.

199 **In November 1960, Robert Galambos figured:** Robert Galambos's biographical details from Squire, 1998.

200 **"Quite possibly the most important roles of glia":** Barres, 2008. This quotation also appears in Martin, 2010. For more on the importance of glia, see Barres, 2008; Wang and Bordey, 2008; Allen, 2009; Edwards, 2009; Sofroniew and Vinters, 2010; Steinhäuser and Seifert, 2010; and Eroglu and Barres, 2010.

201 **"Truth-tellers are genuinely passionate":** Edited from a pre-press edition of Downes and Nunes, 2014. Downes and Nunes interviewed me for this part of their book as an example of a "truth-teller."

205 **In 1960, the Puppeteers' annual Puppetry Festival:** There is a photograph of the event program in the Jim Henson Archive at http://bit.ly/puppetry1960.

205 **Mike and Frances befriended a first-time attendee:** Biographical details about Jim Henson and Frank Oz are mainly from Jones, 2013; Davis, 2009; and the Muppet Wiki at http://bit.ly/muppetwiki.

205 **he wanted to be a journalist, not a puppeteer:** Douglas, 2007.

206 **Henson and Oz found two new Muppets:** The story of Bert and Ernie draws from the Wikipedia entry at http://bit.ly/erniebert.

207 **After the words "In Color," two clay animation:** The first episode of *Sesame Street* can be seen on YouTube at http://bit.ly/firstsesamestreet.

208 **"Bert and Ernie are two grown men sharing a house":** Various sources, including the Muppet Wiki, attribute this quotation to a radio broadcast by Chambers in 1994. See http://bit.ly/gayberternie.

210 **an animated television series they created:** *South Park,* the television series, which first aired in 1997, is based on two animated shorts that Parker and Stone created in 1992 and 1995.

210 **Parker and Stone let filmmaker Arthur Bradford:** *Six Days to Air: The Making of South Park* (2011), sometimes known as *Six Days to South Park,* directed by Arthur Bradford.

214 **In 1998, Viacom asked the two men:** Quotations and details about the making of the *South Park* movie from Pond, 2000.

216 **In 2006, Peter Skillman, an industrial designer:** TED (Technology, Entertainment, Design) 2006. Video at http://bit.ly/skillmanTED.

216 **developed with Dennis Boyle:** Skillman gives details on the genesis of the marshmallow challenge on the TED website at http://bit.ly/skillmanbackground.

217 **Creative professional Tom Wujec confirmed this:** Wujec's slides, and a talk he gave at the 2010 TED conference, at http://bit.ly/wujecTED.

217 **"Several teams will have the powerful desire":** From Wujec's marshmallow challenge instructions at http://bit.ly/marshmallowinstructions.

218 **"Although children's use of tools":** Quotations from Vygotsky, 1980.

222 **In 1954, something unprecedented happened:** Cornwell, 2010.

223 **Before microsociology, the dominant assumption:** Model adapted from David McDermott's website Decision Making Confidence at http://bit.ly /mcdermottdavid.

224 **Sociologist Erving Goffman called the moves:** Collins, 2004.

225 **the average office worker attends:** Data from my own online survey of 123 self-described "office workers," working at various levels of their organizations.

225 **the more creative an organization is:** See Mankins et al., 2014.

226 **"I was assigned to work with Bill Mylan":** Johnson, 1990.

226 **In 1966, Philip Jackson:** From Jackson, 1966: "The other curriculum might be described as unofficial or perhaps even hidden, because to date it has received scant attention from educators. This hidden curriculum can also be represented by three R's, but not the familiar one of reading, 'riting, and 'rithmetic. It is, instead, the curriculum of rules, regulations, and routines, of things teachers and students must learn if they are to make their way with minimum pain in the social institution called *the school.*"

226 **"The crowds, the praise, and the power":** Jackson quotations in this section from Jackson, 1968.

228 **"The personal qualities":** The complete quotation from Jackson is:
"The personal qualities that play a role in intellectual mastery are very different from those that characterize the Company Man. Curiosity, as an instance, is of little value in responding to the demands of conformity. The curious person

typically engages in a kind of probing, poking, and exploring that is almost antithetical to the attitude of the passive conformist. The scholar must develop the habit of challenging authority and questioning the value of tradition. He must insist on explanations for things that are unclear. Scholarship requires discipline, to be sure, but this discipline serves the demands of scholarship rather than the wishes and desires of other people. In short, intellectual mastery calls for sublimated forms of aggression rather than for submission to constraints."

229–230 **airplanes killed 2.2 million people:** Wartime casualty figures are notoriously unreliable and always disputed. To quote statistical historian Matthew White (White, 2013): "The numbers that people want to argue about are casualties." Here, 2.2 million is the sum of casualties and losses listed in the Wikipedia entry "Strategic Bombing During World War II" (http://bit.ly/WW2bombing), which reflects the consensus of historians: 60,595 British civilians; 160,000 airmen in Europe; more than 500,000 Soviet civilians; 67,078 French civilians killed by U.S.-U.K. bombing; 260,000 Chinese civilians; 305,000–600,000 civilians in Germany, including foreign workers; 330,000–500,000 Japanese civilians; 50,000 Italians killed by Allied bombing. Adding these numbers together and taking the high end where there are ranges gives a total of 2,197,673. The sources for these numbers (all of which are listed in the entry itself) include Keegan, 1989; Corvisier and Childs, 1994; and White, 2003.

230 **fired three thousand shells for each bomber:** This number is based on the efficiency of German 88mm guns, or "eighty-eights," at destroying Boeing B-17 Flying Fortresses, which was 2,805 shells per bomber destroyed. Westermann, 2011, cited in Wikipedia at http://bit.ly/surfacetoairmissiles.

231 **the Lockheed SR-72:** Demonstrations of the SR-72 could begin in 2018, with initial flights in 2023, and full service in 2030, according to Brad Leland, Lockheed's portfolio manager for air-breathing hypersonic technologies, in Norris, 2013.

CHAPTER 9: GOOD-BYE, GENIUS

232 **whenever he removed his Quaker-style "wide-awake" hat:** Galton recommends the wide-awake in his book *The Art of Travel* (Galton, 1872)—"I notice that old travellers in both hot and temperate countries have generally adopted a scanty 'wide-awake'"—so I have assumed he may have worn one. The wide-awake is also known as the "Quaker hat." Images at http://bit.ly/wideawakehat.

232 **He wrote later that they were "savages":** Comments from Galton, 1872. For example: "Seizing Food—On arriving at an encampment, the natives commonly run away in fright. If you are hungry, or in serious need of anything that they have, go boldly into their huts, take just what you want, and leave fully adequate payment. It is absurd to be over-scrupulous in these cases."

234 **in Britain, for example, an "E3" carcass is "excellent":** From the EC, or EUROP, classification grid in the U.K. Rural Payments Agency's "Beef Carcase Classification Scheme," available at http://bit.ly/carcase.

234 **"The negro race has occasionally, but very rarely":** Galton, 1869.

237 **In 2010, the average person:** According to the Global Burden of Disease 2010

study (Wang, 2013), the world average life expectency is 67.5 for men and 73.3 for women. The unweighted average of these two values is 70.4, which rounds to 70.

238 **"The power of population is indefinitely greater":** Malthus quotations from Malthus, 1798, and subsequent editions.

239 **famine declined as population increased:** See Devereux and Berge, 2000, for a comprehensive study of famine in the twentieth century.

239 **the First and Second World Wars combined:** Data from Pinker, 2010, which uses Brecke, 1999; Long and Brecke, 2003; and McEvedy and Jones, 1978 as sources.

BIBLIOGRAPHY

Adams, Douglas. *Life, the Universe and Everything (Hitchhiker's Guide to the Galaxy)*. Random House Publishing Group. Kindle Edition, 2008.

Albini, Adriana, Francesca Tosetti, Vincent W. Li, Douglas M. Noonan, and William W. Li. "Cancer Prevention by Targeting Angiogenesis." *Nature Reviews Clinical Oncology* 9, no. 9 (2012): 498–509.

Allen, Nicola J., and Ben A. Barres. "Neuroscience: Glia—More Than Just Brain Glue." *Nature* 457, no. 7230 (2009): 675–77.

Allen, Walter Ernest, ed. *Writers on Writing*. Phoenix House, 1948.

Altman, Lawrence K. "A Scientist, Gazing Toward Stockholm, Ponders 'What If?'" *New York Times,* December 6, 2005.

Amabile, Teresa M. *Creativity and Innovation in Organizations*. Harvard Business School Publishing, 1996.

———. *Creativity in Context: Update to "The Social Psychology of Creativity."* Westview Press, 1996.

———. *How to Kill Creativity*. Harvard Business School Publishing, 1998.

———. "Motivating Creativity in Organizations: On Doing What You Love and Loving What You Do." *California Management Review* 40, no. 1 (1997).

———. "Motivational Synergy: Toward New Conceptualizations of Intrinsic and Extrinsic Motivation in the Workplace." *Human Resource Management Review* 3, no. 3 (1993): 185–201.

———. "The Social Psychology of Creativity: A Componential Conceptualization." *Journal of Personality and Social Psychology* 45, no. 2 (1983): 357.

Amabile, Teresa M., Sigal G. Barsade, Jennifer S. Mueller, and Barry M. Staw. "Affect and Creativity at Work." *Administrative Science Quarterly* 50, no. 3 (2005): 367–403.

Amabile, Teresa M., Regina Conti, Heather Coon, Jeffrey Lazenby, and Michael Herron. "Assessing the Work Environment for Creativity." *Academy of Management Journal* 39, no. 5 (1996): 1154–84.

Amabile, Teresa M., Beth A. Hennessey, and Barbara S. Grossman. "Social Influences on Creativity: The Effects of Contracted-for Reward." *Journal of Personality and Social Psychology* 50, no. 1 (1986): 14.

Amabile, Teresa M., Karl G. Hill, Beth A. Hennessey, and Elizabeth M. Tighe. "The Work Preference Inventory: Assessing Intrinsic and Extrinsic Motivational Orientations." *Journal of Personality and Social Psychology* 66, no. 5 (1994): 950.

Antoniou, Anthony, P. D. P. Pharoah, Steven Narod, Harvey A. Risch, Jorunn E. Eyfjord, J. L. Hopper, Niklas Loman, Håkan Olsson, O. Johannsson, and Åke Borg. "Average Risks of Breast and Ovarian Cancer Associated with *BRCA1* or *BRCA2* Mutations Detected in Case Series Unselected for Family History: A Combined Analysis of 22 Studies." *The American Journal of Human Genetics* 72, no. 5, (2003): 1117–30.

Aristotle, *Nicomachean Ethics.* Translated by Robert C. Bartlett and Susan D. Collins. University of Chicago Press, 2011.

Arnold, Lois B. "The Bascom-Goldschmidt-Porter Correspondence: 1907 to 1922." *Earth Sciences History* 12, no. 2 (1993): 196–223.

Ashby, Ross. *Design for a Brain.* John Wiley, 1952.

Authier, André. *Early Days of X-Ray Crystallography.* Oxford University Press, 2013.

Bailey, Brian P., and Joseph A. Konstan. "On the Need for Attention-Aware Systems: Measuring Effects of Interruption on Task Performance, Error Rate, and Affective State." *Computers in Human Behavior* 22, no. 4 (2006): 685–708.

Barres, Ben A. "The Mystery and Magic of Glia: A Perspective on Their Roles in Health and Disease." *Neuron* 60, no. 3 (2008): 430–40.

Barrows, David Prescott. *The Ilongot or Ibilao of Luzon.* Science Press, 1910.

Bartolomei, F., E. Barbeau, M. Gavaret, M. Guye, A. McGonigal, J. Regis, and P. Chauvel. "Cortical Stimulation Study of the Role of Rhinal Cortex in Deja Vu and Reminiscence of Memories." *Neurology* 63, no. 5 (2004): 858–64.

Bates, Brian R. "Coleridge's Letter from a 'Friend' in Chapter 13 of the Biographia Literaria." Paper presented at the Rocky Mountain Modern Language Association (RMMLA) Conference. Boulder, CO, October 11–13, 2012.

Baum, L. Frank. *The Wonderful Wizard of Oz.* Oxford University Press, 2008.

Beckett, Samuel. *Worstward Ho.* John Calder, 1983.

Benfey, O. Theodore. "August Kekule and the Birth of the Structural Theory of Organic Chemistry in 1858." *Journal of Chemical Education* 35 (1958): 21.

Biello, David. "Fact or Fiction?: Archimedes Coined the Term 'Eureka!' in the Bath." Scientific American. December 8, 2006.

Block, Melissa, and Audie Cornish. "Why Woody Allen Is Always MIA at Oscars." NPR. *All Things Considered.* February 24, 2012.

Boland, C. Richard, Guenter Krejs, Michael Emmett, and Charles Richardson. "A Birthday Celebration for John S. Fordtran, MD." *Proceedings (Baylor University Medical Center)* 25, no. 3 (July 2012): 250–53.

Boston Evening Transcript. "Anthropologist Loses Life." March 31, 1909.

Bragg, William. *Concerning the Nature of Things.* 1925. Reprint, Courier Dover Publications, 2004.

Brecke, P. "The Conflict Dataset: 1400 A.D.–Present." Georgia Institute of Technology, 1999.

Brewster, David. *Memoirs of the Life, Writings, and Discoveries of Sir Isaac Newton.* Vol. 2. Edmonston and Douglas, 1860.

Bronowski, Jacob. *The Ascent of Man.* BBC Books, 2013.

Brooke, Alan, and Lesley Kipling. *Liberty or Death: Radicals, Republicans and Luddites c. 1793–1823.* Workers History Publications, 1993.

Brown, Jonathon D., and Frances M. Gallagher. "Coming to Terms with Failure: Private Self-Enhancement and Public Self-Effacement." *Journal of Experimental Social Psychology* 28, no. 1 (1992): 3–22.

Brown, Marcel. "The 'Lost' Steve Jobs Speech from 1983; Foreshadowing Wireless Networking, the iPad, and the App Store." *Life, Liberty, and Technology.* October 2, 2012. http://lifelibertytech.com/2012/10/02/the-lost-steve-jobs-speech-from-1983-foreshadowing-wireless-networking-the-ipad-and-the-app-store/#.

Burks, Barbara, Dortha Jensen, and Lewis Terman. *The Promise of Youth: Follow-up Studies of a Thousand Gifted Children.* Vol. 3 of *Genetic Studies of Genius Volume.* Stanford University Press, 1930.

Burroughs, Edgar Rice. *A Princess of Mars.* 1917. Reprint, eStar Books, 2012.

Burton, Robert. *On Being Certain: Believing You Are Right Even When You're Not.* St. Martin's Griffin, 2009.

Byers, Nina, and Gary Williams. *Out of the Shadows: Contributions of Twentieth-Century Women to Physics.* Cambridge University Press, 2010.

Cain, Susan. *Quiet: The Power of Introverts in a World That Can't Stop Talking.* Broadway Books, 2013.

———. "The Rise of the New Groupthink." *New York Times,* January 13, 2012.

Calhoun, Craig. "Robert K. Merton Remembered." *Footnotes: Newsletter of the American Sociological Society* 31, no. 33 (2003). http://www.asanet.org/footnotes/mar03/indextwo.html.

Callahan, Michael P., Karen E. Smith, H. James Cleaves, Josef Ruzicka, Jennifer C. Stern, Daniel P. Glavin, Christopher H. House, and Jason P. Dworkin. "Carbonaceous Meteorites Contain a Wide Range of Extraterrestrial Nucleobases." *Proceedings of the National Academy of Sciences* 108, no. 34 (2011): 13995–98.

Cameron, Judy, and W. David Pierce. "Reinforcement, Reward, and Intrinsic Motivation: A Meta-Analysis." *Review of Educational Research* 64, no. 3 (1994): 363–423.

Cameron, Ken. *Vanilla Orchids: Natural History and Cultivation.* Timber Press, 2011.

Caneva, Kenneth L. "Possible Kuhns in the History of Science: Anomalies of Incommensurable Paradigms." *Studies in History and Philosophy of Science* 31, no. 1 (2000): 87–124.

Carmi, Shai, Ken Y. Hui, Ethan Kochav, Xinmin Liu, James Xue, Fillan Grady, Saurav Guha, Kinnari Upadhyay, Dan Ben-Avraham, Semanti Mukherjee, B. Monica Bowen, Tinu Thomas, Joseph Vijai, Marc Cruts, Guy Froyen, Diether Lambrechts, Stéphane Plaisance, Christine Van Broeckhoven, Philip Van Damme, Herwig Van Marck, Nir Barzilai, Ariel Darvasi, Kenneth Offit, Susan Bressman, Laurie J. Ozelius, Inga Peter, Judy H. Cho, Harry Ostrer, Gil Atzmon, Lorraine N. Clark, Todd Lencz, and Itsik Pe'er. "Sequencing an Ashkenazi Reference Panel

Supports Population-Targeted Personal Genomics and Illuminates Jewish and European Origins." *Nature Communications* 5, no 4835 (September 9, 2014).

Carruthers, Peter. "The Cognitive Functions of Language." *Behavioral and Brain Sciences* 25, no. 6 (2002): 657–74.

———. "Creative Action in Mind." *Philosophical Psychology* 24, no. 4 (2011): 437–61.

———. "Human Creativity: Its Cognitive Basis, Its Evolution, and Its Connections with Childhood Pretence." *British Journal for the Philosophy of Science* 53, no. 2 (2002): 225–49.

Carruthers, Peter, and Peter K Smith. *Theories of Theories of Mind.* Cambridge University Press, 1996.

Carus-Wilson, E. M. "The English Cloth Industry in the Late Twelfth and Early Thirteenth Centuries." *Economic History Review* 14, no. 1 (1944): 32–50.

———. "An Industrial Revolution of the Thirteenth Century." *Economic History Review* 11, no. 1 (1941): 39–60.

Caselli, R. J. "Creativity: An Organizational Schema." *Cognitive and Behavioral Neurology* 22, no. 3 (2009): 143–54.

Chadwick, David. *Crooked Cucumber: The Life and Zen Teaching of Shunryu Suzuki.* Harmony, 2000.

Christner, Brent C., Cindy E. Morris, Christine M. Foreman, Rongman Cai, and David C. Sands. "Ubiquity of Biological Ice Nucleators in Snowfall." *Science* 319, no. 5867 (2008): 1214.

Chrysikou, Evangelia G. "When a Shoe Becomes a Hammer: Problem Solving as Goal-Derived, Ad Hoc Categorization." PhD Diss., Temple University, 2006.

Cipolla, Carlo M. *Literacy and Development in the West.* Penguin Books, 1969.

Coleridge, Samuel Taylor. *Biographia Literaria.* 2 vols. Oxford University Press, 1907.

"College Aide Ends Life." *New York Times,* February 24, 1940.

Collins, Randall. *Interaction Ritual Chains.* Princeton University Press, 2004.

———. "Interaction Ritual Chains, Power and Property: The Micro-Macro Connection as an Empirically Based Theoretical Problem." *Micro-Macro Link* (1987): 193–206.

———. "On the Microfoundations of Macrosociology." *American Journal of Sociology* (1981): 984–1014.

Coleridge, Samuel Taylor. 2011. *The Complete Poetical Works of Samuel Taylor Coleridge.* Vols. I and II. Kindle Edition. Public domain.

Cooke, Robert. *Dr. Folkman's War: Angiogenesis and the Struggle to Defeat Cancer.* Random House, 2001.

Comarow, Avery. "Best Children's Hospitals 2013–14: The Honor Roll." *U.S. News & World Report,* June 10, 2014.

Cornell University Library, Division of Rare & Manuscript Collections. "How Did Mozart Compose?" 2002. http://rmc.library.cornell.edu/mozart/compose.htm.

———. "The Mozart Myth: Tales of a Forgery." 2002, http://rmc.library.cornell.edu/mozart/myth.htm.

Cornwell, Erin York. "Opening and Closing the Jury Room Door: A Sociohistorical Consideration of the 1955 Chicago Jury Project Scandal." *Justice System Journal* 31, no. 1 (2010): 49–73.

Corvisier, André, and John Childs. A Dictionary of Military History and the Art of War. Wiley-Blackwell, 1994.

Costa, Marta D., Joana B. Pereira, Maria Pala, Verónica Fernandes, Anna Olivieri, Alessandro Achilli, Ugo A Perego, Sergei Rychkov, Oksana Naumova, and Jiri Hatina. "A Substantial Prehistoric European Ancestry Amongst Ashkenazi Maternal Lineages." *Nature Communications* 4 (2013).

Cox, Catherine Morris. *The Early Mental Traits of Three Hundred Geniuses.* Stanford University Press, 1926.

Cramond, Bonnie. "The Torrance Tests of Creative Thinking: From Design Through Establishment of Predictive Validity." In Rena Faye Subotnik and Karen D. Arnold, eds., *Beyond Terman: Contemporary Longitudinal Studies of Giftedness and Talent.* Greenwood Publishing Group, 1994.

Cropley, Arthur J. *More Ways Than One: Fostering Creativity.* Ablex Publishing, 1992.

Crowe, Cameron. "The True Life Confessions of Fleetwood Mac." *Rolling Stone* 235 (1977).

Csikszentmihalyi, Mihaly. *Creativity: Flow and the Psychology of Discovery and Invention.* Harper Perennial, 1996.

———. *Finding Flow: The Psychology of Engagement with Everyday Life.* Masterminds Series. Basic Books, 1998.

———. *Flow: The Psychology of Optimal Experience.* Harper Perennial Modern Classics, 2008.

Curie, Marie. "Radium and the New Concepts in Chemistry." Nobel lecture, 1911. http://www.nobelprize.org/nobel prizes/chemistry/laureates/1911/marie-curie- lecture.html.

Darwin, Charles. *The Variation of Animals and Plants Under Domestication.* John Murray, 1868.

Davis, Michael. *Street Gang: The Complete History of Sesame Street.* Penguin Books, 2009.

De Groot, Adriaan. *Thought and Choice in Chess.* Psychological Studies. Mouton De Gruyter, 1978.

Descartes, René. *The Passions of the Soul.* Hackett, 1989.

Dettmer, Peggy. "Improving Teacher Attitudes Toward Characteristics of the Creatively Gifted." *Gifted Child Quarterly* 25, no. 1 (1981): 11–16.

Devereux, Stephen. *Famine in the Twentieth Century.* Brighton: Institute of Development Studies, 2000.

Dickens, Charles. *A Christmas Carol.* Simon and Schuster, 1843.

Dickens, Charles, and Gilbert Ashville Pierce. *The Writings of Charles Dickens: Life, Letters, and Speeches of Charles Dickens; with Biographical Sketches of the Principal Illustrators of Dicken's Works.* Vol. 30. Houghton, Mifflin and Company, 1894.

Dietrich, Arne, and Riam Kanso. "A Review of Eeg, Erp, and Neuroimaging Studies of Creativity and Insight." *Psychological Bulletin* 136, no. 5 (2010): 822.

Dolan, Kerry A. *Inside Inventionland. Forbes* 178, no. 11 (2006): 70ff.

Dorfman, Jennifer, Victor A. Shames, and John F. Kihlstrom. "Intuition, Incubation, and Insight: Implicit Cognition in Problem Solving." *Implicit Cognition* (1996): 257–96.

Dostoyevsky, Fyodor, and Anna Grigoryevna Dostoyevskaya. *Dostoevsky: Letters and Reminiscences.* Books for Libraries Press, 1971.

Douglas, Edward. "A Chat with Frank Oz." ComingSoon.net. August 10, 2007, http://www.comingsoon.net/news/movienews.php?id=23056.

Douglass, A. E. "The Illusions of Vision and the Canals of Mars." (1907).

Downes, Larry, and Paul Nunes. *Big Bang Disruption: Strategy in the Age of Devastating Innovation.* Portfolio, 2014.

Doyle, Sir Arthur Conan, and the Conan Doyle Estate. "The Complete Sherlock Holmes." 1877. Reprint, Complete Works Collection, 2011.

Drew, Trafton, Karla Evans, Melissa L.-H. Võ, Francine L. Jacobson, and Jeremy M. Wolfe. "Informatics in Radiology: What Can You See in a Single Glance and How Might This Guide Visual Search in Medical Images?" *Radiographics* 33, no. 1 (2013): 263–74.

Drew, Trafton, Melissa L.-H. Võ, and Jeremy M Wolfe. "The Invisible Gorilla Strikes Again: Sustained Inattentional Blindness in Expert Observers." *Psychological Science* 24, no. 9 (2013): 1848–53.

Drews, Frank A. "Profiles in Driver Distraction: Effects of Cell Phone Conversations on Younger and Older Drivers." *Human Factors: The Journal of the Human Factors and Ergonomics Society* 46, no. 4 (2004): 640–49.

Driscoll, Carlos A., Marilyn Menotti-Raymond, Alfred L. Roca, Karsten Hupe, Warren E. Johnson, Eli Geffen, Eric H. Harley, Miguel Delibes, Dominique Pontier, Andrew C. Kitchener, Nobuyuki Yamaguchi, Stephen J. O'Brien, and David W. Macdonald. "The Near Eastern Origin of Cat Domestication." *Science* 317, no. 5837 (2007): 519–23.

Duncker, Karl. *On Problem Solving.* Translated by Lynne S. Lees. *Psychological Monographs* 58 (1945): i–113.

———. "Ethical Relativity? (An Enquiry into the Psychology of Ethics)." *Mind* (1939): 39–57.

———. "The Influence of Past Experience upon Perceptual Properties." *The American Journal of Psychology* (1939): 255–65.

Duncker, Karl, and Isadore Krechevsky. "On Solution-Achievement." *Psychological Review* 46, no. 2 (1939): 176.

Dunnette, Marvin D., John Campbell, and Kay Jaastad. "The Effect of Group Participation on Brainstorming Effectiveness for 2 Industrial Samples." *Journal of Applied Psychology* 47, no. 1 (1963): 30.

Eckberg, Douglas Lee, and Lester Hill Jr. "The Paradigm Concept and Sociology: A Critical Review." *American Sociological Review* (1979): 925–37.

Ecott, Tim. *Vanilla: Travels in Search of the Ice Cream Orchid.* Grove Press, 2005.

Edwards, Robert. "What the Neuron Tells Glia." *Neuron* 61, no. 6 (2009): 811–12.

Ehrenburg, Ilya. *Life of the Automobile.* Serpent's Tail, 1929.

Einstein, Albert. "Fundamental Ideas and Problems of the Theory of Relativity." *Les Prix Nobel 1922* (1923): 482–90.

———. "How I Created the Theory of Relativity." Lecture given in Kyoto, December 14, 1922. Translated by Yoshimasa A. Ono. *Physics Today* 35, no. 8 (1982): 45–47.

Eisen, Cliff, and Simon P. Keefe. *The Cambridge Mozart Encyclopedia.* Cambridge University Press, 2007.

Eisenberger, Naomi I., and M. D. Lieberman. "Why Rejection Hurts: A Common Neural Alarm System for Physical and Social Pain." *Trends in Cognitive Sciences* 8, no. 7 (2004): 294–300.

Eisenberger, Naomi I., and Matthew D. Lieberman. "Why It Hurts to Be Left Out:

The Neurocognitive Overlap Between Physical and Social Pain." In *The Social Outcast: Ostracism, Social Exclusion, Rejection, and Bullying,* edited by Kipling D. Williams, Joseph P. Forgas, and William von Hippel, 109–30. Routledge, 2005.

Eisenberger, Robert, and Judy Cameron. "Detrimental Effects of Reward: Reality or Myth?" *American Psychologist* 51, no. 11 (1996): 1153.

Eisenberger, Robert, W. David Pierce, and Judy Cameron. "Effects of Reward on Intrinsic Motivation—Negative, Neutral, and Positive: Comment on Deci, Koestner, and Ryan (1999)." *Psychological Bulletin* 125, no. 6 (1999): 677–91.

Elias, Scott. *Origins of Human Innovation and Creativity.* Vol. 16, Developments in Quaternary Science. Elsevier, 2012.

Eliot, Thomas Stearns. "Banquet Speech: December 10, 1948." In *Nobel Lectures, Literature 1901–1967,* edited by Horst Frenz. Elsevier Publishing, 1969.

Emerson, Ralph Waldo. *Journals of Ralph Waldo Emerson, with Annotations.* University of Michigan Library, 1909.

Emling, Shelley. *Marie Curie and Her Daughters: The Private Lives of Science's First Family.* Palgrave Macmillan, 2013.

Epstein, Stephan R. "Craft Guilds, Apprenticeship, and Technological Change in Preindustrial Europe." *Journal of Economic History* 58, no. 3 (1998): 684–713.

Ercin, A. Ertug, Maite Martinez Aldaya, and Arjen Y. Hoekstra. "Corporate Water Footprint Accounting and Impact Assessment: The Case of the Water Footprint of a Sugar-Containing Carbonated Beverage." *Water Resources Management* 25, no. 2 (2011): 721–41.

Ergenzinger, Edward R., Jr. "The American Inventor's Protection Act: A Legislative History." *Wake Forest Intellectual Property Law Journal* 7 (2006): 145.

Eroglu, Cagla, and Ben A. Barres. "Regulation of Synaptic Connectivity by Glia." *Nature* 468, no. 7321 (2010): 223–31.

Everly, George S., Jr., and Jeffrey M. Lating. *A Clinical Guide to the Treatment of the Human Stress Response.* Springer Series on Stress and Coping. Springer, 2002.

Feist, Gregory J. *The Psychology of Science and the Origins of the Scientific Mind.* Yale University Press, 2008.

Feldhusen, John F., and Donald J. Treffinger. "Teachers' Attitudes and Practices in Teaching Creativity and Problem-Solving to Economically Disadvantaged and Minority Children." *Psychological Reports* 37, no. 3f (1975): 1161–62.

Fermi, Laura, and Gilberto Bernardini. *Galileo and the Scientific Revolution.* Dover, 2003.

Ferry, Georgina. *Dorothy Hodgkin: A Life.* Cold Spring Harbor Laboratory Press, 2000.

Festinger, Leon. *Conflict, Decision, and Dissonance.* Stanford University Press, 1964.
———. *A Theory of Cognitive Dissonance.* 1957. Reprint, Stanford University Press, 1962.
———. "Cognitive Dissonance." *Scientific American* 207, no. 4 (1962): 92–102.

Festinger, Leon, Kurt W. Back, and Stanley Schachter. *Social Pressures in Informal Groups: A Study of Human Factors in Housing.* Stanford University Press, 1950.

Festinger, Leon, and James M. Carlsmith. "Cognitive Consequences of Forced Compliance." *Journal of Abnormal and Social Psychology* 58, no. 2 (1959): 203.

Festinger, Leon, Henry W. Riecken, and Stanley Schachter. *When Prophecy Fails: A Social and Psychological Study of a Modern Group That Predicted the Destruction of the World.* Harper Torchbooks, 1956.

Fields, Rick. *How the Swans Came to the Lake*. Shambhala, 1992.

Flaherty, Alice Weaver. *The Midnight Disease: The Drive to Write, Writer's Block, and the Creative Brain*. Mariner Books, 2005.

Flournoy, Théodore. *From India to the Planet Mars: A Study of a Case of Somnambulism*. Harper & Bros., 1900.

Flynn, Francis J., and Jennifer A. Chatman. "Strong Cultures and Innovation: Oxymoron or Opportunity." In *International Handbook of Organizational Culture and Climate* edited by Cary L. Cooper, Sue Cartwright, and P. Christopher Earley. Wiley, 2001, 263–87.

Franklin, Rosalind E. "Location of the Ribonucleic Acid in the Tobacco Mosaic Virus Particle." *Nature* 177, no. 4516 (1956): 929–30.

——. "Structural Resemblance Between Schramm's Repolymerised A-Protein and Tobacco Mosaic Virus." *Biochimica et Biophysica Acta* 18 (1955): 313–14.

——. "Structure of Tobacco Mosaic Virus." *Nature* 175, no. 4452 (1955): 379.

Franklin, Rosalind E., Donald L. D. Caspar, and Aaron Klug. "The Structure of Viruses as Determined by X-Ray Diffraction." *Plant Pathology, Problems and Progress* 1958 (1958): 447–61.

Franklin, Rosalind E., and Barry Commoner. "Abnormal Protein Associated with Tobacco Mosaic Virus; X-Ray Diffraction by an Abnormal Protein (B8) Associated with Tobacco Mosaic Virus." *Nature* 175, no. 4468 (1955): 1076.

Franklin, Rosalind E., and A. Klug. "The Nature of the Helical Groove on the Tobacco Mosaic Virus Particle X-Ray Diffraction Studies." *Biochimica et Biophysica Acta* 19 (1956): 403–16.

——. "The Splitting of Layer Lines in X-Ray Fibre Diagrams of Helical Structures: Application to Tobacco Mosaic Virus." *Acta Crystallographica* 8, no. 12 (1955): 777–80.

Freedberg, A. Stone, and Louis E. Barron. "The Presence of Spirochetes in Human Gastric Mucosa." *American Journal of Digestive Diseases* 7, no. 10 (1940): 443–45.

Freeman, Karen. "Dr. Ian A. H. Munro, 73, Editor of the Lancet Medical Journal." *New York Times,* February 3, 1997.

Fuller, Steve. *Thomas Kuhn: A Philosophical History for Our Times*. University of Chicago Press, 2001.

Gabai-Kapara, Efrat, Amnon Lahad, Bella Kaufman, Eitan Friedman, Shlomo Segev, Paul Renbaum, Rachel Beeri, Moran Gal, Julia Grinshpun-Cohen, Karen Djemal, Jessica B. Mandell, Ming K. Lee, Uziel Beller, Raphael Catane, Mary-Claire King, and Ephrat Levy-Lahad. "Population-Based Screening for Breast and Ovarian Cancer Risk Due to *BRCA1* and *BRCA2*." *Proceedings of the National Academy of Sciences,* September 5, 2014.

Galilei, Galileo. "La Bilancetta." *Galileo and the Scientific Revolution* (1961): 133–43.

Galton, Francis. *The Art of Travel; or, Shifts and Contrivances Available in Wild Countries*. (1872). Digitized June 29, 2006. Google Book.

——. *English Men of Science: Their Nature and Nurture*. D. Appleton, 1875.

——. *Hereditary Genius*. Macmillan, 1869.

——. *Inquiries into Human Faculty and Its Development*. Macmillan, 1883.

——. *Natural Inheritance*. Macmillan, 1889.

Gardner, Howard E. *Creating Minds: An Anatomy of Creativity Seen Through the Lives of Freud, Einstein, Picasso, Stravinsky, Eliot, Graham, and Ghandi*. Basic Books, 2011.

Garfield, Eugene. "A Different Sort of Great-Books List—the 50 20th-Century Works Most Cited in the Arts and Humanities Citation Index, 1976–1983." *Current Contents* 16 (1987): 3–7.

Getzels, Jacob W., and Philip W. Jackson. *Creativity and Intelligence: Explorations with Gifted Students.* Wiley, 1962, pp. xvii, 293.

Glasser, William. *Positive Addiction.* Harper & Row New York, 1976.

———. "Positive Addiction." *Journal of Extension* (May/June 1977): 4–8.

———. "Promoting Client Strength Through Positive Addiction." *Canadian Journal of Counselling and Psychotherapy/Revue Canadienne de Counseling et de Psychothérapie* 11, no. 4 (2012).

Gleick, James. "The Paradigm Shifts." *New York Times Magazine,* December 29, 1996.

Glucksberg, Sam. "The Influence of Strength of Drive on Functional Fixedness and Perceptual Recognition." *Journal of Experimental Psychology* 63, no. 1 (1962): 36.

Glynn, Jenifer. *My Sister Rosalind Franklin: A Family Memoir.* Oxford University Press, 2012.

Gonzales, Laurence. *Deep Survival: Who Lives, Who Dies, and Why.* W. W. Norton, 2004.

Gould, Stephen Jay. *Ever Since Darwin.* W. W. Norton, 1977.

Guralnick, Peter. *Searching for Robert Johnson.* Dutton Adult, 1989.

Hadamard, Jacques. *The Mathematician's Mind.* Princeton University Press, 1996.

Haines, Catharine M. C. *International Women in Science: A Biographical Dictionary to 1950.* ABC-CLIO, 2001.

Ham, Denise. *Marie Sklodowska Curie: The Woman Who Opened the Nuclear Age.* 21st Century Science Associates, 2002.

Hansen, Amy. "Lyrics of Rap and Lines of Stage Help Mattapan Teen Turn to Better Life." *Boston Globe,* December 5, 2012.

Harbluk, Joanne L., Y. Ian Noy, and Moshe Eizenman. *The Impact of Cognitive Distraction on Driver Visual Behaviour and Vehicle Control.* Transport Canada, 2002., http://www.tc.gc.ca/motorvehiclesafety/tp/tp13889/pdf/tp13889es.pdf.

Harlow, Harry F. "Learning and Satiation of Response in Intrinsically Motivated Complex Puzzle Performance by Monkeys." *Journal of Comparative and Physiological Psychology* 43, no. 4 (1950): 289.

———. "The Nature of Love." *American Psychologist* 13, no. 12 (1958): 673.

Harlow, Harry F., Margaret Kuenne Harlow, and Donald R. Meyer. "Learning Motivated by a Manipulation Drive." *Journal of Experimental Psychology* 40, no. 2 (1950): 228.

Hegel, Georg Wilhelm Friedrich. *The Philosophy of History.* Translated by J. Sibree. Courier Dover Publications, 2004.

Heidegger, Martin. *The Principle of Reason.* Studies in Continental Thought. Indiana University Press, 1956.

Heider, F. *The Psychology of Interpersonal Relations.* Psychology Press, 1958.

Heilman, Kenneth M. *Creativity and the Brain.* Psychology Press, 2005.

Hélie, Sebastien, and Ron Sun. "Implicit Cognition in Problem Solving." In *The Psychology of Problem Solving: An Interdisciplinary Approach,* edited by Sebastien Helie. Nova Science Publishing, 2012.

———. "Incubation, Insight, and Creative Problem Solving: A Unified Theory and a Connectionist Model." *Psychological Review* 117, no. 3 (2010): 994.

Hennessey, B. A., and T. M. Amabile. "Creativity." *Annual Review of Psychology* 61 (2010): 569–98.

Hennessey, Beth A., and Teresa M. Amabile. *Creativity and Learning (What Research Says to the Teacher).* National Education Association, 1987.

Heppenheimer, T. A. *First Flight: The Wright Brothers and the Invention of the Airplane.* Wiley, 2003.

Hill, John Spencer. *A Coleridge Companion: An Introduction to the Major Poems and the "Biographia Literaria."* Prentice Hall College Division, 1984.

Hollander, Jason. "Renowned Columbia Sociologist and National Medal of Science Winner Robert K. Merton Dies at 92." Columbia News: The Public Affairs and Record Home Page, February 5, 2003, http://www.columbia.edu/cu/news/03/02 /robertKMerton.html.

Holmes, Chris E., Jagoda Jasielec, Jamie E. Levis, Joan Skelly, and Hyman B. Muss. "Initiation of Aspirin Therapy Modulates Angiogenic Protein Levels in Women with Breast Cancer Receiving Tamoxifen Therapy." *Clinical and Translational Science* 6, no. 5 (2013): 386–90.

Holmes, K. C., and Rosalind E. Franklin. "The Radial Density Distribution in Some Strains of Tobacco Mosaic Virus." *Virology* 6, no. 2 (1958): 328–36.

Hope, Jack. "A Better Mousetrap." *American Heritage,* October 1996, 90–97.

Hornaday, Anna. "Woody Allen on 'Rome,' Playing Himself and Why He Skips the Oscars." *Washington Post,* June 28, 2012.

Hume, David. *An Enquiry Concerning Human Understanding.* Oxford Philosophical Texts. Oxford University Press, 1748.

Huxley, Aldous. *Texts and Pretexts: An Anthology with Commentaries.* 1932. Reprint, Greenwood, 1976.

Hyman, Ira E., S. Matthew Boss, Breanne M. Wise, Kira E. McKenzie, and Jenna M. Caggiano. "Did You See the Unicycling Clown? Inattentional Blindness While Walking and Talking on a Cell Phone." *Applied Cognitive Psychology* 24, no. 5 (2010): 597–607.

Isherwood, Christopher. *Goodbye to Berlin.* HarperCollins, 1939.

Ito, S. "Anatomic Structure of the Gastric Mucosa." *Handbook of Physiology* 2 (1967): 705–41.

Iyer, Pico. "The Joy of Quiet." *New York Times,* December 29, 2011.

Jackson, Philip W. *Life in Classrooms.* Teachers College Press, 1968.

———. "The Student's World." *Elementary School Journal* (1966): 345–57.

Jahn, Otto. *Life of Mozart.* 3 vols. Cambridge Library Collection: Music. Cambridge University Press, 2013.

Associated Press. "Jap Reception Center Nears Completion." April 4, 1942. University of California Japanese American Relocation Digital Archive Photograph Collection. http://bit.ly/japreception.

Jenkin, John. *William and Lawrence Bragg, Father and Son: The Most Extraordinary Collaboration in Science.* Oxford University Press, 2008.

Johnson, Clarence L. "Kelly," with Maggie Smith. *Kelly: More Than My Share of It All.* Random House, 1990.

Jones, Brian Jay. *Jim Henson: The Biography.* Ballantine Books, 2013.

Jørgensen, Jes K., Cécile Favre, Suzanne E. Bisschop, Tyler L. Bourke, Ewine F. van Dishoeck, and Markus Schmalzl. "Detection of the Simplest Sugar, Glycolaldehyde, in a Solar-Type Protostar with Alma." *Astrophysical Journal Letters* 757, no. 1 (2012): L4.

Kahn, Roger. *The Boys of Summer.* Harper Perennial Modern Classics, 1972.

Kahneman, Daniel. *Thinking, Fast and Slow.* Farrar, Straus and Giroux, 2013.

———. "Attention and Effort." Prentice-Hall, 1973.

———. "Don't Blink! The Hazards of Confidence." *New York Times,* October 23, 2011.

Kahneman, Daniel, and Gary Klein. "Conditions for Intuitive Expertise: A Failure to Disagree." *American Psychologist* 64, no. 6 (2009): 515.

Kahneman, Daniel, and Amos Tversky. "Choices, Values, and Frames." *American Psychologist* 39, no. 4 (1984): 341.

———. "On the Psychology of Prediction." *Psychological Review* 80, no. 4 (1973): 237.

———. "Subjective Probability: A Judgment of Representativeness," *Cognitive Psychology* 3, no. 3 (1972): 430–54.

Kaufman, James C., and Robert J. Sternberg, eds. *The Cambridge Handbook of Creativity* Cambridge Handbooks in Psychology. Cambridge University Press, 2010.

Keegan, John. *The Second World War.* Random House, 1989.

Kenning, Kaleene. "Ohabai Shalome Synagogue." Examiner.com, March 2, 2010.

Kepler, Johannes. *The Six-Cornered Snowflake.* Paul Dry Books, 1966.

Kidd, Mark, and Irvin M. Modlin. "A Century of *Helicobacter pylori.*" *Digestion* 59, no. 1 (1998): 1–15.

Kimble, Gregory A., and Michael Wertheimer. *Portraits of Pioneers in Psychology.* Vol. 3. American Psychological Association, 1998.

King, Stephen. *Danse Macabre.* Gallery Books, 2010.

———. *On Writing: A Memoir of the Craft.* Pocket Books, 2001.

Kleinmuntz, Benjamin. *Formal Representation of Human Judgment.* Carnegie Series on Cognition. John Wiley & Sons, 1968.

Kraybill, Donald B., Karen M. Johnson-Weiner, and Steven M. Nolt. *The Amish.* Johns Hopkins University Press, 2013.

Kroger, S., B. Rutter, R. Stark, S. Windmann, C. Hermann, and A. Abraham. "Using a Shoe as a Plant Pot: Neural Correlates of Passive Conceptual Expansion." *Brain Research* 1430 (2012): 52–61.

Kruger, J., and D. Dunning. "Unskilled and Unaware of It: How Difficulties in Recognizing One's Own Incompetence Lead to Inflated Self-Assessments." *Journal of Personality and Social Psychology* 77, no. 6 (1999): 1121–34.

Krützen, Michael, Janet Mann, Michael R. Heithaus, Richard C. Connor, Lars Bejder, and William B. Sherwin. "Cultural Transmission of Tool Use in Bottlenose Dolphins." *Proceedings of the National Academy of Sciences of the United States of America* 102, no. 25 (2005): 8939–43.

Kuhn, Thomas S. *Black-Body Theory and the Quantum Discontinuity, 1894–1912.* University of Chicago Press, 1987.

———. *The Copernican Revolution: Planetary Astronomy in the Development of Western Thought.* Harvard University Press, 1992.

———. *The Essential Tension: Selected Studies in Scientific Tradition and Change.* University of Chicago Press, 1977.

———. *The Road Since Structure: Philosophical Essays, 1970–1993, with an Autobiographical Interview.* University of Chicago Press, 2002.

———. *The Structure of Scientific Revolutions.* 3rd ed. University of Chicago Press, 1996.

Lach-Szyrma, Wladyslaw Somerville. *Aleriel; or, A Voyage to Other Worlds. A Tale, Etc.* 1883. Reprint, British Library, Historical Print Editions, 2011.

Lakatos, Imre, Thomas S. Kuhn, W. N. Watkins, Stephen Toulmin, L. Pearce Williams, Margaret Masterman, and P. K. Feyerabend. *Criticism and the Growth of Knowledge: Proceedings of the International Colloquium in the Philosophy of Science, London, 1965.* Cambridge University Press, 1970.

Langer, J. S. "Instabilities and Pattern Formation in Crystal Growth." *Reviews of Modern Physics* 52, no. 1 (1980): 1.

Langton, Christopher G., and Katsunori Shimohara. *Artificial Life V: Proceedings of the Fifth International Workshop on the Synthesis and Simulation of Living Systems (Complex Adaptive Systems).* A Bradford Book, 1997.

Laplace, Pierre Simon. *Essai Philosophique sur les Probabilités.* Vve Courcier, 1814.

Largo, Michael. *God's Lunatics: Lost Souls, False Prophets, Martyred Saints, Murderous Cults, Demonic Nuns, and Other Victims of Man's Eternal Search for the Divine.* William Morrow Paperbacks, 2010.

Lawson, Carol. *Behind the Best Sellers: Stephen King.* Westview Press, 1979.

Lax, Eric. *Woody Allen: A Biography.* Da Capo Press, 2000.

——. *Conversations with Woody Allen.* Random House, 2009.

Lee, R. E., Jr., Gareth J. Warren, and Lawrence V. Gusta. *Biological Ice Nucleation and Its Applications.* American Phytopathological Society, 1995.

Lehrer, Jonah. *Imagine: How Creativity Works.* Houghton Mifflin, 2012.

Lennard, Reginald. *Rural England, 1086–1135: A Study of Social and Agrarian Conditions.* Oxford: Clarendon Press, 1959.

——. "Early English Fulling Mills: Additional Examples." *Economic History Review* 3, no. 3 (1951): 342–43.

Libbrecht, Kenneth G. "Morphogenesis on Ice: The Physics of Snow Crystals." *Engineering and Science* 64, no. 1 (2001): 10–19.

Linde, Nancy. *Cancer Warrior.* First aired on February 27, 2001, by PBS. Written, produced and directed by Nancy Linde.

Lindsay, Kenneth C., and Peter Vergo, eds. *Kandinsky: Complete Writings on Art.* Da Capo Press, 1994.

Long, William J., and Peter Brecke. *War and Reconciliation: Reason and Emotion in Conflict Resolution.* MIT Press, 2003.

Lowell, Percival. "Tores of Saturn." *Lowell Observatory Bulletin* 1 (1907): 186–90.

Lucas, Adam. *Wind, Water, Work: Ancient and Medieval Milling Technology.* Leiden, Koninklijke Brill, 2006.

Lum, Timothy E., Rollin J. Fairbanks, Elliot C. Pennington, and Frank L. Zwemer. "Profiles in Patient Safety: Misplaced Femoral Line Guidewire and Multiple Failures to Detect the Foreign Body on Chest Radiography." *Academic Emergency Medicine* 12, no. 7 (2005): 658–62.

Mack, Arien, and Irvin Rock. *Inattentional Blindness.* A Bradford Book, 2000.

MacLeod, Hugh. *Ignore Everybody: And 39 Other Keys to Creativity.* Portfolio Hardcover, 2009.

Maddox, Brenda. *Rosalind Franklin: The Dark Lady of DNA.* Harper Perennial, 2003.

Malthus, Thomas Robert. *An Essay on the Principle of Population, as It Affects the Future Improvement of Society.* Dent, 1973.

Mankins, Michael, Chris Brahm, and Gregory Caimi. "Your Scarcest Resource." *Harvard Business Review* 92, no. 5 (2014): 74–80.

Mann, Thomas. *Deutsche Ansprache: Ein Appell an die Vernunft (An Appeal to Reason)*. S. Fischer, 1930.

Markel, Howard. "'I Have Seen My Death': How the World Discovered the X-Ray." PBS NewsHour: The Rundown, December 20, 2012, http://www.pbs.org /newshour/rundown/i-have-seen-my-death-how-the-world-discovered-the-x-ray.

Marshall, Barry. *Helicobacter Pioneers: Firsthand Accounts from the Scientists Who Discovered Helicobacters, 1892–1982*. Wiley-Blackwell, 2002.

Marshall, Barry J., and J. Robin Warren. "Unidentified Curved Bacilli in the Stomach of Patients with Gastritis and Peptic Ulceration." *Lancet* 323, no. 8390 (1984): 1311–15.

Martin, Douglas. "Robert Galambos, Neuroscientist Who Showed How Bats Navigate, Dies at 96." *New York Times,* July 15, 2010.

Massey, Howard. *Behind the Glass: Top Record Producers Tell How They Craft the Hits*. Backbeat Books, 2000.

Masterman, Margaret. "The Nature of a Paradigm." In Lakatos, Imre, and Alan Musgrave, eds. *Criticism and the Growth of Knowledge*. Cambridge University Press, 1970.

McEvedy, Colin, and Richard Jones. *Atlas of World Population History*. Harmondsworth: Penguin Books, 1978.

McGraw, Kenneth O., and John C. McCullers. "Evidence of a Detrimental Effect of Extrinsic Incentives on Breaking a Mental Set." *Journal of Experimental Social Psychology* 15, no. 3 (1979): 285–94.

Merton, Robert K. *On the Shoulders of Giants: A Shandean Postscript*. University of Chicago Press, 1993.

——. "The Matthew Effect in Science." *Science* 159, no. 3810 (1968): 56–63.

——. "The Matthew Effect in Science, II: Cumulative Advantage and the Symbolism of Intellectual Property." *Isis* (1988): 606–23.

Metcalfe, Janet, and David Wiebe. "Intuition in Insight and Noninsight Problem Solving." *Memory & Cognition* 15, no. 3 (1987): 238–46.

Meyer, Steven J. "Introduction: Whitehead Now." *Configurations* 13 (2005): 1–33.

Mithen, Steven, ed. *Creativity in Human Evolution and Prehistory*. Routledge, 2014.

——. *The Prehistory of the Mind: The Cognitive Origins of Art, Religion and Science*. Thames & Hudson, 1996.

Momsen, Bill. "*Mariner IV:* First Flyby of Mars: Some Personal Experiences." http://bit .ly/billmomsen (2006).

Morozov, Evgeny. *To Save Everything, Click Here: The Folly of Technological Solutionism*. PublicAffairs, 2013.

Morris, James M. *On Mozart*. Woodrow Wilson Center Press and Cambridge University Press, 1994.

Mossberg, Walt. "The Steve Jobs I Knew." *AllThingsD,* October 5, 2012. http:// allthingsd.com/20121005/the-steve-jobs-i-knew/.

Moszkowski, Alexander. *Conversations with Einstein*. Horizon Press, 1973.

Mueller, Jennifer S., Shimul Melwani, and Jack A. Goncalo. "The Bias Against Creativity: Why People Desire but Reject Creative Ideas." *Psychological Science* 23, no. 1 (2012): 13–17.

Munro, Ian. "Pyloric Campylobacter Finds a Volunteer." *Lancet* 1, no. 8436 (1985): 1021–22.

———. "Spirals and Ulcers." *Lancet* 1, no. 8390 (1984): 1336–37.

Munro, John. "The Symbiosis of Towns and Textiles: Urban Institutions and the Changing Fortunes of Cloth Manufacturing in the Low Countries and England, 1270–1570." *Journal of Early Modern History* 3, no. 3 (1999): 1–74.

Munro, John H. "Industrial Energy from Water-Mills in the European Economy, 5th to 18th Centuries: The Limitations of Power." University Library of Munich, 2002.

Nakaya, Ukichiro. *Snow Crystals: Natural and Artificial.* Harvard University Press, 1954.

Neisser, Ulric, and Nicole Harsch. "Phantom Flashbulbs: False Recollections of Hearing the News About *Challenger.*" In *Affect and Accuracy in Recall: Studies of "Flashbulb" Memories,* edited by Eugene Winograd and Ulric Neisser. Cambridge University Press, 1992, pp. 9–31.

Newell, Allen, J. Clifford Shaw, and Herbert Alexander Simon. *The Processes of Creative Thinking.* Rand Corporation, 1959.

Newton, Isaac, I. Bernard Cohen, and Marie Boas Hall. *Isaac Newton's Papers & Letters on Natural Philosophy and Related Documents.* Harvard University Press, 1978.

Nickles, Thomas. *Thomas Kuhn.* Contemporary Philosophy in Focus. Cambridge University Press, 2002.

Nisbett, Richard E., and Timothy D. Wilson. "Telling More Than We Can Know: Verbal Reports on Mental Processes." *Psychological Review* 84, no. 3 (1977): 231.

Norris, Guy. *Skunk Works Reveals SR-71 Successor Plan.* New York: Springer-Verlag, 2013.

Ochse, R. A. *Before the Gates of Excellence: The Determinants of Creative Genius.* Cambridge Greek and Latin Classics. Cambridge University Press, 1990.

Ogburn, William F., and Dorothy Thomas. 1922. "Are Inventions Inevitable? A Note on Social Evolution." *Political Science Quarterly* 37, no. 1 (March 1922): 83–98.

Olton, Robert M. "Experimental Studies of Incubation: Searching for the Elusive." *Journal of Creative Behavior* 13, no. 1 (1979): 9–22.

Olton, Robert M., and David M. Johnson. "Mechanisms of Incubation in Creative Problem Solving." *American Journal of Psychology* (1976): 617–30.

Osborn, Alex F. *Applied Imagination: Principles and Procedures of Creative Problem-Solving.* C. Scribner's Sons, 1957.

———. *How to Think Up.* McGraw-Hill, 1942.

Paine, Thomas. *The Age of Reason.* 1794. Reprint, CreateSpace Independent Publishing Platform, 2008.

———. *Writings of Thomas Paine: (1779–1792), The Rights of Man.* Vol. 2. 1791. Reprint, 2013.

Pareto, Vilfredo, Arthur Livingston, Andrew Bongiorno, and James Harvey Rogers. *A Treatise on General Sociology.* General Publishing Company, 1935.

Pearson, Barry Lee, and Bill McCulloch. *Robert Johnson: Lost and Found.* Music in American Life. University of Illinois Press, 2003.

Pelham, R. A. "The Distribution of Early Fulling Mills in England and Wales." *Geography* (1944): 52–56.

Penn, D. C., K. J. Holyoak, and D. J. Povinelli. "Darwin's Mistake: Explaining the Discontinuity Between Human and Nonhuman Minds." *Behavioral and Brain Sciences* 31, no. 2 (2008): 109–30.

Penrose, Roger, and Martin Gardner. *Emperors New Mind: Concerning Computers, Minds, and the Laws of Physics.* Oxford University Press, 1989.

Pincock, Stephen. "Nobel Prize Winners Robin Warren and Barry Marshall." *Lancet* 366, no. 9495 (2005): 1429.

Pinker, Steven. *The Better Angels of Our Nature: Why Violence Has Declined*. Viking, 2010.

Plath, Slyvia. *Journals of Sylvia Plath*. Dial Press, 1982.

Pollio, Marcus Vitruvius. *The Ten Books on Architecture*. Architecture Classics, 2013.

Pond, Steve. "Trey Parker and Matt Stone: The *Playboy* Interview." *Playboy* 457, no. 7230 (2000): 675–77.

Porter, Mary Winearls. *What Rome Was Built With: A Description of the Stones Employed in Ancient Times for Its Building and Decoration*. University of Michigan Library, 1907.

Price, Monica T. "The Corsi Collection in Oxford." Corsi Collection of Decorative Stones. Oxford University Museum website: http://www.oum.ox.ac.uk/corsi /about/oxford.

Pynchon, Thomas. "Is It O.K. to Be a Luddite?" *New York Times*, October 28, 1984.

Radack, David V. "Getting Inventorship Right the First Time." *JOM* 46, no. 6 (1994): 62.

Rakauskas, Michael E., Leo J. Gugerty, and Nicholas J. Ward. "Effects of Naturalistic Cell Phone Conversations on Driving Performance." *Journal of Safety Research* 35, no. 4 (2004): 453–64.

Ramsey, E. J., K. V. Carey, W. L. Peterson, J. J. Jackson, F. K. Murphy, N. W. Read, K. B. Taylor, J. S. Trier, and J. S. Fordtran. "Epidemic Gastritis with Hypochlorhydria." *Gastroenterology* 76, no. 6 (1979): 1449–57.

Read, J. Don, and Darryl Bruce. "Longitudinal Tracking of Difficult Memory Retrievals." *Cognitive Psychology* 14, no. 2 (1982): 280–300.

Read, Leonard E. "I, Pencil." *Freeman*, December 1958, p. 32.

Renfrew, Colin and Iain Morley. *Becoming Human: Innovation in Prehistoric Material and Spiritual Culture*. Cambridge University Press, 2009.

Rensberger, Boyce. "David Krech, 68, Dies; Psychology Pioneer." *New York Times*, July 16, 1977.

Rich, Ben R., and Leo Janos. *Skunk Works: A Personal Memoir of My Years at Lockheed*. Little Brown, 1994.

Richardson, John. *A Life of Picasso*. Vol. 2, *1907–1917: The Painter of Modern Life*. Random House, 1996.

Rietzschel, Eric F., Bernard A. Nijstad, and Wolfgang Stroebe. "The Selection of Creative Ideas After Individual Idea Generation: Choosing Between Creativity and Impact." *British Journal of Psychology* 101, no. 1 (2010): 47–68.

Rosaldo, Michelle Zimbalist. *Knowledge and Passion*. Cambridge University Press, 1980.

Rothenberg, Albert. "Creative Cognitive Processes in Kekule's Discovery of the Structure of the Benzene Molecule." *American Journal of Psychology* (1995): 419–38.

Rubright, Linda. "D.Inc.tionary." *Medium*. February 16, 2013. https://medium.com /@deliciousday/d-inc-tionary-b8eed806fc6b.

Runco, Mark A. "Creativity Has No Dark Side." In *The Dark Side of Creativity*, edited by David H. Cropley, Arthur J. Cropley, James C. Kaufman, and Mark A. Runco. Cambridge University Press, 2010.

Rutter, B., S. Kroger, H. Hill, S. Windmann, C. Hermann, and A. Abraham. "Can Clouds Dance? Part 2: An Erp Investigation of Passive Conceptual Expansion." *Brain and Cognition* 80, no. 3 (2012): 301–10.

"S.F. Clear of All But 6 Sick Japs." *San Francisco Chronicle.* May 21, 1942, http://www
.sfmuseum.org/hist8/evac19.html.

Sawyer, R. Keith. *Explaining Creativity: The Science of Human Innovation.* Oxford University Press, 2012.

Schnall, Simone. *Life as the Problem: Karl Duncker's Context.* Psychology Today Tapes, 1999.

Schrodinger, Erwin. *What Is Life?: With Mind and Matter and Autobiographical Sketches.* Canto Classics. Cambridge University Press, 1944.

Seale, Jack. "The Joy of Difficult Second (or Third, or Twelfth) Albums." *Radio Times,* May 17, 2012.

Seger, Carol A. "How Do the Basal Ganglia Contribute to Categorization? Their Roles in Generalization, Response Selection, and Learning Via Feedback." *Neuroscience & Biobehavioral Reviews* 32, no. 2 (2008): 265–78.

Semmelweis, Ignaz. *The Etiology, Concept, and Prophylaxis of Childbed Fever.* 1859. Reprint, University of Wisconsin Press, 1983.

Senzaki, Nyogen. *101 Zen Stories.* Kessinger Publishing, 1919.

Sheehan, William. *The Planet Mars: A History of Observation and Discovery.* University of Arizona Press, 1996.

Sheehan, William, and Thomas Dobbins. "The Spokes of Venus: An Illusion Explained." *Journal for the History of Astronomy* 34 (2003): 53–63.

Sheh, Alexander, and James G Fox. "The Role of the Gastrointestinal Microbiome in *Helicobacter pylori* Pathogenesis." *Gut Microbes* 4, no. 6 (2013): 22–47.

Shepardson, George Defrees. *Electrical Catechism: An Introductory Treatise on Electricity and Its Uses.* McGraw-Hill, 1908.

Shurkin, Joel N. *Terman's Kids: The Groundbreaking Study of How the Gifted Grow Up.* Little Brown, 1992.

Sidowski, J. B., and D. B. Lindsley. "Harry Frederick Harlow: October 31, 1905–December 6, 1981." *Biographical Memoirs of the National Academy of Sciences* 58 (1988): 219–57.

Simon, Herbert A. *Karl Duncker and Cognitive Science.* Springer, 1999.

Simon, Herbert A., Allen Newell, and J. C. Shaw. "The Processes of Creative Thinking." Rand Corporation, 1959.

Simonton, Dean Keith. *Greatness: Who Makes History and Why.* Guilford Press, 1994.

———. *Origins of Genius: Darwinian Perspectives on Creativity.* Oxford University Press, 1999.

Simpson, Eileen. *Poets in Their Youth: A Memoir.* Noonday Press, 1982.

Smithgall, Elsa, ed. *Kandinsky and the Harmony of Silence: Painting with White Border (Phillips Collection).* Yale University Press, 2011.

Snyder, Thomas D. *120 Years of American Education: A Statistical Portrait.* National Center for Education Statistics, 1993.

Sofroniew, Michael V. and Harry V. Vinters. "Astrocytes: Biology and Pathology." *Acta Neuropathologica* 119, no. 1 (2010): 7–35.

Spinoza, Benedictus de. *Ethics: Ethica Ordine Geometrico Demonstrata.* 1677. Reprint, Floating Press, 2009.

Squire, Larry R. *The History of Neuroscience in Autobiography.* Vol. 1. Academic Press, 1998.

Staw, Barry M. "Why No One Really Wants Creativity." *Creative Action in Organizations* (1995): 161–66.

Steinhäuser, Christian, and Gerald Seifert. "Astrocyte Dysfunction in Temporal Lobe Epilepsy." *Epilepsia* 51, no. s5 (2010): 54–54.

Strauss, David. "Percival Lowell, W. H. Pickering and the Founding of the Lowell Observatory." *Annals of Science* 51, no. 1 (1994): 37–58.

Strayer, David L., and Frank A. Drews. "Cell-Phone-Induced Driver Distraction." *Current Directions in Psychological Science* 16, no. 3 (2007): 128–31.

Strayer, David L., Frank A. Drews, and Dennis J. Crouch. "A Comparison of the Cell Phone Driver and the Drunk Driver." *Human Factors: The Journal of the Human Factors and Ergonomics Society* 48, no. 2 (2006): 381–91.

Strayer, David L., Frank A. Drews, and William A. Johnston. "Cell Phone–Induced Failures of Visual Attention During Simulated Driving." *Journal of Experimental Psychology: Applied* 9, no. 1 (2003): 23.

Suzuki, Shunryu. *Zen Mind, Beginner's Mind.* 1970. Reprint, Shambhala, 2011.

Syrotuck, William and Syrotuck, Jean Anne. *Analysis of Lost Person Behavior.* Barkleigh Productions, 2000.

Takeuchi, H., Y. Taki, H. Hashizume, Y. Sassa, T. Nagase, R. Nouchi, and R. Kawashima. "The Association Between Resting Functional Connectivity and Creativity." *Cereb Cortex* 22, no. 12 (2012): 2921–29.

——. "Cerebral Blood Flow During Rest Associates with General Intelligence and Creativity." *PLoS One* 6, no. 9 (2011): e25532.

Taylor, Frederick Winslow. *The Principles of Scientific Management.* Harper, 1911.

Terman, Lewis. *Genetic Studies of Genius.* Vol. 1. Stanford Press, 1925.

——. *Genetic Studies of Genius.* Vol. 5. Stanford Press, 1967.

——. *Sex and Personality Studies in Masculinity and Femininity.* Shelley Press, 2007.

——. "Are Scientists Different?" *Scientific American* 192 (1955): 25–29.

——. *Condensed Guide for the Stanford Revision of the Binet-Simon Intelligence Tests.* Nabu Press, 2010.

——. *Genius and Stupidity: A Study of Some of the Intellectual Processes of Seven "Bright" and Seven "Stupid" Boys.* The Pedagogical Seminary 13, no. 3 (1906).

——. *The Intelligence of School Children: How Children Differ in Ability, the Use of Mental Tests in School Grading and the Proper Education of Exceptional Children.* Riverside Textbooks in Education. Houghton Mifflin Company, 1919.

Terman, Lewis M., and M. A. Merrill. *Stanford-Binet Intelligence Scale.* Houghton Mifflin Company, 1960.

Terman, Lewis Madison, and Maud A. Merrill. *Measuring Intelligence: A Guide to the Administration of the New Revised Stanford-Binet Tests of Intelligence.* Riverside Textbooks in Education. Houghton Mifflin, 1937.

Terman, Lewis, and Melita Oden. *The Gifted Child Grows Up: Twenty-Five Years' Follow-up of a Superior Group.* Vol. 4 of *Genetic Studies of Genius.* Stanford University Press, 1947.

Terman, Lewis M., and Melita H. Oden. *The Gifted Group at Mid-Life.* Stanford University Press, 1959.

Torrance, Ellis Paul. *Norms Technical Manual: Torrance Tests of Creative Thinking.* Ginn, 1974.

————. "The Creative Personality and the Ideal Pupil." *Teachers College Record* 65, no. 3 (1963): 220–26.

Towne, Henry R. "Engineer as Economist." *Transactions of the American Society of Mechanical Engineers* 7, no. 1886 (1886): 425ff. Reprinted in *Academy of Management Proceedings* vol. 1986, no. 1: 3–4.

Trabert, Britton, Roberta B. Ness, Wei-Hsuan Lo-Ciganic, Megan A. Murphy, Ellen L. Goode, Elizabeth M. Poole, Louise A. Brinton, Penelope M. Webb, Christina M. Nagle, and Susan J. Jordan. "Aspirin, Nonaspirin Nonsteroidal Anti-Inflammatory Drug, and Acetaminophen Use and Risk of Invasive Epithelial Ovarian Cancer: A Pooled Analysis in the Ovarian Cancer Association Consortium." *Journal of the National Cancer Institute* 106, no. 2 (2014): djt431.

Truzzi, Marcello. "On the Extraordinary: An Attempt at Clarification." *Zetetic Scholar* 1, no. 11 (1978).

Tsoref, Daliah, Tony Panzarella, and Amit Oza. "Aspirin in Prevention of Ovarian Cancer: Are We at the Tipping Point?" *Journal of the National Cancer Institute* 106, no. 2 (2014): djt453.

Tsu, Lao. *Tao Te Ching.* Vintage Books, 1972.

Tversky, Amos, and Daniel Kahneman. "Advances in Prospect Theory: Cumulative Representation of Uncertainty." *Journal of Risk and Uncertainty* 5, no. 4 (1992): 297–323.

————. "Availability: A Heuristic for Judging Frequency and Probability." *Cognitive Psychology* 5, no. 2 (1973): 207–32.

————. "The Framing of Decisions and the Psychology of Choice." *Science* 211, no. 4481 (1981): 453–58.

————. "Judgment Under Uncertainty: Heuristics and Biases." *Science* 185, no. 4157 (1974): 1124–31.

————. "Loss Aversion in Riskless Choice: A Reference-Dependent Model." *Quarterly Journal of Economics* 106, no. 4 (1991): 1039–61.

————. "Rational Choice and the Framing of Decisions." *Journal of Business* (1986): S251–78.

Tyson, Neil deGrasse. "The Perimeter of Ignorance." *Natural History* 114, no. 9 (2005).

————. "The Perimeter of Ignorance." Talk adapted from *Natural History Magazine,* given at the *Beyond Belief: Science, Religion, Reason and Survival* Conference. Salk Institute for Biological Studies, La Jolla, California, November 5, 2006. Video at http://bit.ly/NdGTSalk.

Underwood, Geoffrey D. M. *Implicit Cognition.* Oxford University Press, 1996.

Unge, Peter. "*Helicobacter pylori* Treatment in the Past and in the 21st Century." In *Helicobacter Pioneers: Firsthand Accounts from the Scientists Who Discovered Helicobacter,* edited by Barry Marshall. Wiley, 2002, 203–13.

United States Presidential Commission on the Space Shuttle Challenger Accident. *Report to the President: Actions to Implement the Recommendations of the Presidential Commission on the Space Shuttle Challenger Accident.* National Aeronautics and Space Administration, 1986.

Vallerand, Robert J. "On the Psychology of Passion: In Search of What Makes People's Lives Most Worth Living." *Canadian Psychology/Psychologie Canadienne* 49, no. 1 (2008): 1.

Vallerand, Robert J., Céline Blanchard, Genevieve A. Mageau, Richard Koestner, Catherine Ratelle, Maude Léonard, Marylene Gagné, and Josée Marsolais. "Les Passions de l'Âme: On Obsessive and Harmonious Passion." *Journal of Personality and Social Psychology* 85, no. 4 (2003): 756.

Vallerand, Robert J., and Nathalie Houlfort. "Passion at Work." In *Emerging Perspectives on Values in Organizations,* edited by Stephen W. Gilliland, Dirk D. Steiner, and Daniel P. Skarlicki. Information Age Publishing, 2003, 175–204.

Vallerand, Robert J., Yvan Paquet, Frederick L. Philippe, and Julie Charest. "On the Role of Passion for Work in Burnout: A Process Model." *Journal of Personality* 78, no. 1 (2010): 289–312.

Vallerand, Robert J., Sarah-Jeanne Salvy, Geneviève A. Mageau, Andrew J. Elliot, Pascale L. Denis, Frédéric M. E. Grouzet, and Celine Blanchard. "On the Role of Passion in Performance." *Journal of Personality* 75, no. 3 (2007): 505–34.

Valsiner, Jaan, ed. *Thinking in Psychological Science: Ideas and Their Makers.* Transaction Publishers, 2007.

Van Der Weyden, Martin B., Ruth M. Armstrong, and Ann T. Gregory. "The 2005 Nobel Prize in Physiology or Medicine." *Medical Journal of Australia* 183, nos. 11–12 (2005): 612.

Vernon, P. E., ed. *Creativity: Selected Readings.* Penguin Books, 1970.

Vul, Edward, and Harold Pashler. "Incubation Benefits Only After People Have Been Misdirected." *Memory & Cognition* 35, no. 4 (2007): 701–10.

Vygotsky, Lev S. *Mind in Society: The Development of Higher Psychological Processes.* Harvard University Press, 1980.

Wald, Elijah. *Escaping the Delta: Robert Johnson and the Invention of the Blues.* Amistad, 2004.

Wallace, Alfred Russel. *Is Mars Habitable? A Critical Examination of Professor Percival Lowell's Book "Mars and Its Canals," with an Alternative Explanation.* Macmillan, 1907.

———. *Man's Place in the Universe: A Study of the Results of Scientific Research in Relation to the Unity or Plurality of Worlds.* Chapman and Hall, 1904.

Wallace, David Foster. *This Is Water: Some Thoughts, Delivered on a Significant Occasion, About Living a Compassionate Life.* Little, Brown, 2009.

Wallas, Graham. *The Art of Thought.* Harcourt, Brace, 1926.

Wang, Doris D. and Angélique Bordey. "The Astrocyte Odyssey." *Progress in Neurobiology* 86, no. 4 (2008): 342–67.

Wang, Haidong, Laura Dwyer-Lindgren, Katherine T. Lofgren, Julie Knoll Rajaratnam, Jacob R. Marcus, Alison Levin-Rector, Carly E. Levitz, Alan D. Lopez, and Christopher J. L. Murray. "Age-Specific and Sex-Specific Mortality in 187 Countries, 1970–2010: A Systematic Analysis for the Global Burden of Disease Study 2010." *The Lancet* 380, no. 9859 (2013): 2071–94.

Wardlow, Gayle Dean. *Chasin' That Devil Music: Searching for the Blues.* Backbeat Books, 1998.

Warren, Robin J. "Helicobacter: The Ease and Difficulty of a New Discovery." Nobel lecture, December 8, 2005. http://www.nobelprize.org/nobel_prizes/medicine /laureates/2005/warren-lecture.pdf.

Weide, Robert B. *Woody Allen: A Documentary.* PBS "American Masters" documentary originally aired on television in 2011. Video on Demand release dated 2013.

Weinberg, Steven. "The Revolution That Didn't Happen." *New York Review of Books* 25, no. 3 (1998): 250–53.

Weisberg, Robert, and Jerry M. Suls. "An Information-Processing Model of Duncker's Candle Problem." *Cognitive Psychology* 4, no. 2 (1973): 255–76.

Weisberg, Robert W. *Creativity: Beyond the Myth of Genius.* W. H. Freeman, 1993.

———. *Creativity: Genius and Other Myths.* Series of Books in Psychology. W. H. Freeman, 1986.

———. *Creativity: Understanding Innovation in Problem Solving, Science, Invention, and the Arts.* Wiley, 2006.

———. "On the 'Demystification' of Insight: A Critique of Neuroimaging Studies of Insight." *Creativity Research Journal* 25, no. 1 (2013): 1–14.

———. "Toward an Integrated Theory of Insight in Problem Solving." *Thinking & Reasoning.* February 24, 2014, http://www.tandfonline.com/doi/abs/10.1080/13546783 .2014.886625#.U9u-vYBdW6i.

Weisskopf-Joelson, Edith, and Thomas S. Eliseo. "An Experimental Study of the Effectiveness of Brainstorming." *Journal of Applied Psychology* 45, no. 1 (1961): 45.

Werrell, Kenneth P. "The Strategic Bombing of Germany in World War II: Costs and Accomplishments." *Journal of American History* 73, no. 3 (December 1986): 702–13.

Westby, Erik L., and V. L. Dawson. "Creativity: Asset or Burden in the Classroom?" *Creativity Research Journal* 8, no. 1 (1995): 1–10.

Westermann, Edward B. *Flak: German Anti-Aircraft Defenses, 1914–1945.* University Press of Kansas, 2001.

White, Matthew. *Atrocities: The 100 Deadliest Episodes in Human History.* W. W. Norton, 2013.

———. *Historical Atlas of the Twentieth Century.* Matthew White, 2003, http://users.erols .com/mwhite28/20centry.htm.

Whitehead, Alfred North. *Religion in the Making: Lowell Lectures 1926.* 1926. Reprint, Fordham University Press, 1996.

Whitson, Jennifer A., and Adam D. Galinsky. "Lacking Control Increases Illusory Pattern Perception." *Science* 322, no. 5898 (2008): 115–17.

Wolfram, Stephen. "The Personal Analytics of My Life." Stephen Wolfram blog, March 8, 2012. http://bit.ly/wolframanalytics.

Wozniak, Steve, with Gina Smith. *iWoz: Computer Geek to Cult Icon; How I Invented the Personal Computer, Co-Founded Apple, and Had Fun Doing It.* W. W. Norton, 2007.

Wright, Orville, and Wilbur Wright. *The Early History of the Airplane.* Dayton-Wright Airplane Company, 1922. https://archive.org/details/earlyhistoryofai00wrigrich.

Young, Kristie, Michael Regan, and M. Hammer. "Driver Distraction: A Review of the Literature." In *Distracted Driving,* 379–405. Australasian College of Road Safety, 2007.

Yule, Sarah S. B. *Borrowings: A Collection of Helpful and Beautiful Thoughts.* New York: Dodge Publishing, 1889.

Zaslaw, Neal. "Recent Mozart Research and *Der neue Köchel.*" In *Musicology and Sister Disciplines: Past, Present, Future. Proceedings of the 16th International Congress of the International Musicological Society, London, 1997.* Oxford University Press, 2000.

———. "Mozart as a Working Stiff." Essay published on the "Apropos Mozart" web site, 1994. http://bit.ly/zaslaw.

Zemlo, Tamara R., Howard H. Garrison, Nicola C. Partridge, and Timothy J. Ley. "The Physician-Scientist: Career Issues and Challenges at the Year 2000." *FASEB Journal* 14, no. 2 (2000): 221–30.

Zepernick, Bernhard, and Wolfgang Meretz. "Christian Konrad Sprengel's Life in Relation to His Family and His Time: On the Occasion of His 250th Birthday." *Willdenowia-Annals of the Botanic Garden and Botanical Museum Berlin-Dahlem* 31, no. 1 (2001): 141–52.

Zuckerman, Harriet. *Scientific Elite: Nobel Laureates in the United States.* Transaction Publishers, 1977.

Zuckerman, Harriet Anne. *Nobel Laureates in the United States: A Sociological Study of Scientific Collaboration.* PhD diss. Columbia University, 1965.

INDEX

Apple Computer, 47
Apple Inc., 45–46, 47, 48, 50, 99
Archimedes, 38–39
Aristotle, 86, 106, 107, 262*n*
Armstrong, Daisy, 265*n*
Army Air Forces (USAAF), U.S., 203
art:
  from angst, 169–70
  creation of, 56–59
  music and, 169–70, 174–75, 183–84
artificial intelligence, 16
Ashby, Ross, 15
Ashkenazi Jews, 138, 269*n,* 270*n*
Aspen, Colo., 47
astrocytes, 200
astronomy, 111–13, 263*n,* 264*n*
Atlanta, Ga., 144, 145, 147
Auble, Pamela, 37
Auschwitz, 30
Austin, Tex., 143, 144, 145
Austin Chronicle, 167
Australia, 91–93, 143–44, 184
authorship, 6
  as misleading, 9, 122, 123–24, 125
  records of, 6–7
  see also credit; inventors; patents
automation, 143, 158, 161
  see also looms, automatic
Avedon, Richard, 70
awards, *see* rewards
Aztecs, 2

Bach, Johann Sebastian, 187
Bacon, Francis, 146
bacteria, 137, 153, 260*n*
  Helicobacter pylori, 93, 94, 95, 107
  in stomachs, 91–95, 98, 113
  ulcers and, 92, 93, 95
  viruses vs., 119, 121
Balkan War, First, 57
Bascom, Florence, 133–35, 137
Basilica of San Domenico, 6
Baumgarten, Marjorie, 167
bauxite, 143, 155
Beatles, 169
Becher, Johann Joachim, 262*n*
Beckett, Samuel, 69
beginner's mind, 104–5, 110, 113
Beijerinck, Martinus, 119
Belarus, 217
Belgium, 2–3

beliefs, 66, 116
Belle-Vue, Réunion, 3
Bellier-Beaumont, Elvire, 3
Bellier-Beaumont, Ferréol, 3–4, 5–6,
  17–18, 23, 24
Bellow, Saul, 70
Bergman, Ingmar, 178
Bergman, Torbern, 152
Berlin, Germany, 26–27, 29, 30, 56
Berlin, University of, 27
Berryman, John, 167
Bert and Ernie (Muppets), 207–9, 210,
  222
bicycles, 52, 53, 55
Big Bird (Muppet), 207
"Bilancetta, La" (Galileo), 39
Billboard, 169
bin Laden, Osama, 231
Biographia Literaria (Coleridge), 40
Birkbeck College, 121
Blackbird surveillance plane, 195
Black Death, 239
Blau, Marietta, 126–27
Block, Harry (char.), 177
blood, 74
  tumors and, 60–63
blood vessels, 61–62, 64, 65
Bloomsbury, England, xv
blues, 174–75
Bode, Johann Elert, 263*n*
Boeing, 195
Bogart, Humphrey, 20
Book of Mormon, The (musical), 209–10
Boston, Mass., 182–83
Boston Children's Hospital, 63, 87–88
Boston City Hospital, 62
Boston Globe, 184
Botticelli, Sandro, xv
bottles, bottling, 151–52, 154
Box Problem (Candle Problem), 32, 35–37,
  172
Boyle, Dennis, 216
Boyle, Robert, 132, 137
Bradford, Arthur, 210, 212
Bragg, William, Jr., 133, 134, 135–36, 137
Bragg, William, Sr., 133, 135
brain, 14–15, 30, 86, 114, 199–200
  inattentional blindness and, 96–97
  see also mind
brainstorming, 49–51
BRCA genes, 138, 269*n*–70*n*
Brecht, Bertolt, 26

British *Journal of Cancer,* 62
Brock, David, xvii
Brooklyn Dodgers, 185
Brünn, Czech Republic, 120
Brussels, Belgium, 118
Bryn Mawr College, 133, 134
Buckingham, Lindsey, 169
Burbank, Calif., 194
Burcham, Milo, 193
Burroughs, Edgar Rice, 111, 264*n*
Burton, Robert, 130

caffeine, 145, 156
Caine Mutiny, The (film), 20
Cajal, Santiago Ramón y, 199–200
California, 18, 19, 105, 154, 168–69, 205
California, University of, at Berkeley, 30,
    43, 105
Caligula, Emperor of Rome, 157
Cambridge University, 122, 123
    Girton College, 123
    Newnham College, 123
Canby, Vincent, 167
cancer, 61, 62–63, 65, 138, 269*n*
Candle Problem (Box Problem), 32, 35–37,
    172
cans, canning:
    aluminum in, 143–44, 145, 155
    of Coca-Cola, 143–44, 145–46, 154, 155
    of food, 154
    origin of, 154
    tin in, 154, 155
capitalists, 142
caramel coloring, 144–45
carbon dioxide, 145, 152
Carhart, Suzanne, 267*n*
cars, 146–47
Carter, Shawn Corey (Jay-Z), 184
Cartwright, William, 140–41, 142–43, 158,
    159–61
Caselli, Richard, 13
Cell Biology, 62
certainty, 113–14, 116–17
    false, 114–16
Cézanne, Paul, 146–47
Chagall, Marc, 56
Challenger (space shuttle), 114, 265*n*
Chambers, Joseph, 208
Chandler, Asa, 153
Chanel No. 5, 5
Charlie Problem, 33

Charlotte, N.C., 208
Chartres, Bernard of, 130, 268*n*
chemotherapy, 62
Chicago, University of, 222
children:
    language and, 218–19
    team work of, 216–17, 220–22
Children's Music Foundation, 184
China, 5, 151
chromosomes, 121, 122–23
    origin of word, 120
chromosome theory, 120
cinnamon, 145
City College of New York, 205
Civil War, U.S., 154
Claudius, Emperor of Rome, 157
Clinton, Bill, 78
Coca-Cola, 143–46, 149, 151, 153–54, 155,
    156
    bottling of, 153–54
    cans of, 143–44, 145–46, 154, 155
    ingredients in, 5, 144–45
    as medicine, 153
    as product of the world, 146
    syrup of, 144–45
    water in, 156
Coca-Coca Bottling Company, 144, 145
Coca-Cola Company, 5, 144, 145, 156
cocaine, 145, 152
coca leaf, 145, 153
Codd, Hiram, 153
"cognitive revolution," 30
Coleridge, Samuel Taylor, 39–41, 251*n*
Collins, Randall, 224
Colorado, University of, 42, 209
"Come On Eileen," 170
commitment, 175–76
compliance, creation vs., 227–28
computers, xvi–xvii, 14–15, 46–47
    as originally conceived, xvii
confidence, 117
Copernicus, Nicolaus, 263*n*
Copyright Office, 8
copyrights, 8–9
    increasing numbers of, 8
Cori, Gerty, 124
Cork Problem, 32
Cosby, Bill, 187
Cosby Show, The, 187
creation, creativity, xiii–xviii, 6–25, 26–27,
    30, 32, 64–65
    as action, 225

creation, creativity (*cont'd*)
  in art, 56–59
  attention and, 102, 105
  commitment and, 175–76, 178–79
  compliance vs., 227–28
  consequences of, 150, 236, 237
  education's suppressing of, 18, 84
  emotional pressures and, 169–70
  evaluation and, *see* evaluation
  as execution, 51
  failures and, *see* failures
  faith and, 66
  first steps of, 25, 44–45, 186–87
  genius as prerequisite for, 9, 14, 15–16,
    19, 21–22, 23
  ideas and, 48–49, 51
  as incremental, 13–14, 37, 45, 51, 54,
    57–58, 59, 64, 89
  as innate and ordinary, 13, 14, 17, 18, 23,
    24, 236
  innovation and, 12–13
  insight and, 33, 39–42
  inspiration and, 38–39, 42, 64, 188
  intelligence vs., 84
  interruption and, 188, 190
  language and, 217, 218, 219–20
  in medicine, *see* medicine
  as monotonous, 64–65, 190
  motivation and, *see* motivation
  myths about, xiv–xvi, 8, 13, 18, 24, 25,
    48–49, 90
  patents for, *see* patents
  population growth and, 237, 240
  as problem solving, 16, 17
  rejection and, *see* rejection
  as result of thinking, 31, 37–38, 42, 102
  rewards and, *see* rewards
  in science, *see* science
  selection and, 189–91
  solitude as requirement for, 187, 188
  studies on, 22–23, 83–85
  temptation and, 176, 186
  time and, 24–25, 70–72, 187
  tool chains and, 145, 146–48, 161
  ubiquity of, 10–11
  uniqueness and, 192
  as unwelcome, 74, 76
  valuing of, 84–85
  work and, 24–25, 70–71
  see also technology
creative organizations, 195, 201, 203–4,
    214, 215, 229

  compliance vs. creation in, 227–29
  cooperation and, 220
  hierarchy and, 222
  individual accomplishments and, 220
  interactions within, 223–24
  marshmallow challenge and, 216–17,
    220–21
  planning and, 221, 225–26
  problem solving and, 220
  rituals in, 225
  talking vs. acting in, 225
  see also partnerships; Skunk Works
Creativity: Selected Readings (Vernon), xiv
credit, 7, 9, 122–31
  claiming of, *see* authorship
  for women, 122, 123–24, 125, 126–27
Crick, Francis, 122, 125, 126
Crookes, William, 132, 137
"Cross Road Blues," 175
crystallography, 120–21, 125, 126, 131,
    133–35, 136
  of biological samples, 121, 126, 136–37
  X-ray, 133, 135, 136–37
crystals, 120, 121, 132, 133–34, 136, 267n
Ctesibius, 7
Curie, Irène, 124
Curie, Marie, 123–24, 209
Curie, Pierre, 123, 124, 209
Curiosity (Mars rover), 137
Curse of the Jade Scorpion, The (film), 163
Czech Republic, 120

D-21, 231
Dachau, 29
Darwin, Charles, 3, 24, 120, 233, 238,
    266n–67n
David, Larry, 167
da Vinci, Leonardo, xv, 14
Davison Design, 78–79, 257n–58n
Dayton, Ohio, 52, 54
Dean, Jimmy, 206
Deconstructing Harry, 177
Deep Blue, xvii
Deep South, 174
de Groot, Adriaan, 99–102
Demetrios, Saint, 6
Descartes, René, 182
Design for a Brain (Ashby), 15
Dexys Midnight Runners, 170
"Dialogue Between Franklin and the
    Gout" (Franklin), 157

high-fructose corn syrup, 144, 145, 155–56
Himba, 232–33
Hippocrates, 74
hirudotherapy, 74
Hitchcock, Alfred, 43
Hitler, Adolf, 27, 30, 194, 235
Hodgkin, Dorothy, 135, 136, 137
Holiday, Billie, 17
Hollywood, 169
Holmes, Oliver Wendell, 256n
Holmes, Sherlock (char.), 34–35, 98, 102,
    251n
Hooke, Robert, 130, 131–32
Hooker, William C., 77
Hopeless Diamond, 229, 230, 231
Hopkins, Samuel, 247n–48n
How to Think Up (Osborn), 49
Hume, David, 75
humorism, 74, 75, 129
Huxley, Aldous, 98–99

IBM, 47, 49
ideas, 48–50
    abandoning of, 191
    brainstorming and, 49–51
    new, 64, 74, 83, 85, 87, 199, 200–201, 229
    rejection of, *see* rejection
IDEO, 216
Ilongots, 180–81, 182
I Love Lucy (TV show), 20
Imagine (Lehrer), xiv
inattentional blindness, 95–98, 113
incubation (unconscious thinking),
    42–44, 252n
India, 3, 155
Indiana, 50
Indonesia, 5
Industrial Revolution, 14
information revolution, 160
Ingber, Donald, 63–64, 65
innovation, 12–13, 37, 191
    as series of failures, 63, 65, 69
insight (aha!/eureka moment), 33, 39–42,
    43
inspiration, 38–39, 42, 64, 188
insulin, 136
Intel, 21
intellectual security, 197–98
interactions, 223–24
International Design Conference Aspen
    (IDCA), 253n

Internet, xvii–xviii
"Internet of Things," xvii–xviii
interruption, 188
invention, *see* creation, creativity
Inventionland, 78–79
invention promotion companies, 78–79
inventors, 6
    increasing numbers of, 7–8
    see also specific inventors
iPad, 212
iPhone, 45–46, 99, 252n, 253n
Irene Taylor Trust, 184
Isherwood, Christopher, 27–28, 250n
Israel, 184
Italy, 7
Ito, Susumo, 95
iTunes, 211

Jackson, Phillip, 83–84, 226–28, 275n
Jacobson, Oliver, 183, 184
Jahn, Otto, xiv
Japan, 103, 202
Japanese Americans, 103, 261n
Java, 3
Jay-Z, *see* Carter, Shawn Corey (Jay-Z)
Jefferson, Thomas, 2, 5
Jennifer (tumor patient), 60–61, 64, 65,
    254n
jet engines, 194, 195, 202–3
Jews, Ashkenazi, 138, 269n, 270n
Jobs, Steve, 45, 46, 47, 48, 50, 104
Johns Hopkins University, 133
Johnson, Clarence "Kelly," 194–97, 198,
    201, 202, 203–4, 225–26, 229, 230–31,
    235
Johnson, Robert, 174–75, 176
Johnson, Virginia, 175
Jones, William, 181

Kahn, Roger, 185
Kandinsky, Wassily, 56–59, 69, 104
Kasparov, Garry, xvii
Kasuga, Teruro, 261n
Kaufman, Charlie, 186
Kekulé, August, 41–42
Kenya, 5
Kepler, Johannes, 131, 132, 268n
Kermit the Frog (Muppet), 205, 206
Keynes, John Maynard, 238
Kiev, Ukraine, 6

step-by-step process of, 35, 36–37, 45, 54, 69
in teams, 220
thinking and, 30–36, 43–44, 45, 99–101
"Processes of Creative Thinking, The" (Newell), 15–16
Procter & Gamble, xv–xvi, xvii
propellers, 55, 202
Psychological Review, 28
puerperal fever, 72, 75, 256n
Puppeteers of America, 205
puppets, 205
  see also Muppets
Pure Food and Drug Act, 153
Pushkin, Aleksandr, xiv
Pynchon, Thomas, 87

race, 234–35
radiation, 62
rap, 183, 184
Read, Leonard, 147, 148
Reichelt, Franz, 80–82, 88–89, 92, 258n
Reinhardt, Max, 26
rejection, 72, 73, 79, 88, 90
  fear of, 86
  as information, 88
  of new ideas, 82–83, 85–87
Renaissance, xv, 6, 7, 9, 14
Réunion, 1, 3–4, 5, 23–24
rewards:
  as destructive external influence, 166, 167
  effects of, on creativity, 173–74
  motivation and, 164, 171–73
  open-minded thinking and, 173
  passion and, 181
  quality of work and, 164
Rhinow Hills, Germany, 52
Rich, Ben, 229, 230–31
Richard, Jean Michel Claude, 5–6
Riecken, Henry, 115
Rietzschel, Eric, 85
Rights of Man, The (Paine), 142
Rioch, David, 199, 200, 201
Robinson, Ken, 18
Rock, Irvin, 95
Roman Empire, 157
Romanticism, 14
Rome, ancient, 236
Romeo and Juliet (Shakespeare), 183
Röntgen, Wilhelm, 132, 137

Roosevelt, Theodore, 157
Rosaldo, Michelle, 181
Rostock, Germany, 202
Roth, David Lee, 211
Rousseau, Jean-Jacques, 238
Rowland, Kevin, 170
Rowlf (Muppet), 206
Roxbury, Mass., 182–83
Royal Institution, 135
Royal Perth Hospital, 91–93
Royal School of Mines, 134
Royal Society, 123
Rubright, Linda, 69
Rumours (album), 169, 170, 173
Russia, 28, 29, 57
Russky Viestnik, 171, 173
Rutan, Burt, 216

Sagan, Carl, xv, 75
Sahlin, Don, 206
Saigon, Vietnam, 229
Sainte-Suzanne, Réunion, 1, 3
St. Thomas's Abbey, 120
Salisbury, John of, 130
Salk Institute, 108
San Bruno, Calif., 103
San Domenico, Basilica of, 6
San Francisco, Calif., 103, 105, 154, 261n
San Francisco International Airport, 103
Sarma, Sanjay, xvii
Sausalito, Calif., 168–69
Schachter, Stanley, 115
Schiaparelli, Giovanni, 264n
Schrödinger, Erwin, 118, 126, 266n
Schweppe, Johann Jacob, 152
Schweppes Company, 152
science:
  airplanes and, 52–55
  astronomy and, 111–13, 263n, 264n
  building blocks of, 130, 137–38
  constraints in, 110
  creation in, 8
  men credited for women's work in, 122, 123–24, 125, 126–27, 129
  neuro-, 199–200
  as oppressor of women, 123, 133
  parachutes and, 80–82, 88–89
  paradigms and, 106, 129
  recognition given in, 127–28
  in series of revolutions, 106, 107
  sociology of, 129